Math Intervention

Building Number Power

with Formative Assessments, Differentiation, and Games

Grades PreK-2

Jennifer Taylor-Cox, Ph.D.

EYE ON EDUCATION
6 DEPOT WAY WEST, SUITE 106
LARCHMONT, NY 10538
(914) 833-0551
(914) 833-0761 fax
www.eyeoneducation.com

Library of Congress Cataloging-in-Publication Data

Taylor-Cox, Jennifer.
Math intervention : building number power with formative
assessments, differentiation, and games, PreK-2 / Jennifer Taylor-Cox.
    p. cm.
ISBN 978-1-59667-108-9
1. Mathematics--Study and teaching (Early childhood)
2. Number concept in children.
3. Numeracy. I. Title.
    QA135.6.T3565 2008
    372.7--dc22

                                        2008046897

Book design services provided by
Jennifer Osterhouse Graphic Design
3752 Danube Drive, Davidsonville, MD 21035
(410-798-8585)

Also available from Eye On Education

Math Intervention:
Building Number Power with Formative Assessments,
Differentiation, and Games, Grades 3-5
Jennifer Taylor-Cox

Family Math Night: Math Standards in Action
Jennifer Taylor-Cox

Family Math Night: Middle School Math Standards in Action
Jennifer Taylor-Cox and Christine Oberdorf

Mathematics Coaching Handbook:
Working with Teachers to Improve Instruction
Pia M. Hansen

Differentiated Instruction for K-8 Math and Science:
Activities and Lesson Plans
Mary Hamm and Dennis Adams

A Collection of Performance Tasks and Rubrics:
Primary School Mathematics
Charlotte Danielson and Pia Hansen Powell

Upper Elementary School Mathematics
Charlotte Danielson

Middle School Mathematics
Charlotte Danielson

High School Mathematics
Charlotte Danielson and Elizabeth Marquez

Mathematics and Multi-Ethnic Students: Exemplary Practices
Yvelyne Germain-McCarthy & Katharine Owens

Assessment in Middle and High School Mathematics:
A Teacher's Guide
Daniel Brahier

Teaching Mathematics in the Block
Susan Gilkey and Carla Hunt

Acknowledgements

This book is dedicated to Joey Cox.
Thanks for all of the hours spent playing math games with me.
I enjoyed every minute.

• • • • • • • • • • • • •

Appreciation is extended to friend and colleague, Christine Oberdorf,
an exemplary mathematics educator and incredible staff development teacher.
Thank you for your brain power and your friendship.

Gratitude is expressed to friend and colleague, Karren Schultz-Ferrell, an expert
in early childhood mathematics. Thank you for your generous feedback.

Recognition is extended to friend and colleague, Sheila Etzkorn, an amazing
math educator. You are an inspiration to students and teachers.

Special thanks to friend and colleague, Shirley Sneed, an excellent staff
development teacher. Thanks for your knowledge and support.

Thanks to other reviewers Alice Gabbard, Brenda Love,
Gary Martin, and Virginia Newlin.

Tons of appreciation is offered to Jennifer Osterhouse, a brilliant graphic designer.
It is my honor and pleasure to work with you.

Gratitude is expressed to Marida Hines, a talented illustrator.
Thank you for your artistic perspective.

Extensive acknowledgment and gratitude are offered to the thousands of
students, educators, and parents who have influenced my life. It is for
you and because of you that this essential work is done.

Believe in the power of math!
Jennifer Taylor-Cox

About the Author

 Dr. Jennifer Taylor-Cox is an enthusiastic, captivating presenter and well-known educator. She is the Executive Director of **Taylor-Cox Instruction, LLC**. Jennifer serves as a consultant in mathematics education providing professional development opportunities for numerous districts across the United States. Her keynote speeches, workshops, seminars, classroom demonstration lessons, and presentations are always high-energy and insightful.

Jennifer earned her Ph.D. from the University of Maryland and was awarded the "Outstanding Doctoral Research Award" from the University of Maryland and the "Excellence in Teacher Education Award" from Towson University. She served as the president of the Maryland Council of Teachers of Mathematics. Jennifer is the author of many professional articles and books including *Family Math Night: Math Standards in Action* and *Family Math Night: Middle School Math Standards in Action*.

Dr. Taylor-Cox knows how to make learning mathematics engaging and fun which motivates learners of all ages! She has a passion for differentiated math instruction. Jennifer believes that all students are entitled to targeted instruction aimed at their specific learning needs. Jennifer lives and has her office in Severna Park, Maryland. She is the mother of three children. Jennifer truly understands how to connect research and practice in education. Her zeal for mathematics education is alive in her work with educators, students, and parents.

If you are interested in learning more about the professional development opportunities Dr. Taylor-Cox offers, please feel free to contact her at **Taylor-Cox Instruction, LLC**.

Jennifer Taylor-Cox, Ph.D., Educational Consultant
Office: 410-729-5599, Fax: 410-729-3211
Email: Jennifer@Taylor-CoxInstruction.com

Free Downloads

The 31 games displayed in this book are also available on Eye On Education's Web site as Adobe Acrobat PDF files. Those who have purchased this book have permission to download them and print them out. You can access these downloads by visiting Eye On Education's Web site: **www.eyeoneducation.com**.

Click on **FREE Downloads** or search or browse our Web site to find this book and then scroll down for downloading instructions.

You'll need your book-buyer access code: MAI-7108-9

List of Downloads

Table of Contents

Introduction

What is Math Intervention?

Math intervention is targeted instruction for students struggling with mathematics. Too often students have difficulty understanding math concepts, and the instruction they receive perpetuates or masks the problems rather than addressing them. Students are left to flail and founder in a confusing sea of procedures and rules. The problems become more profound as the student moves into each new grade. As math educators we can no longer let this happen. We must make math meaningful for students and provide targeted instruction that focuses on the precise academic needs of students. It is essential that we concentrate on repairing students' misconceptions and learning gaps in ways that build their math capacity.

In mathematics there are four interconnected goals; accuracy, efficiency, flexibility, and fluency. Our first priority is to help students with their math accuracy, which means we help them learn how to obtain the correct answer. Next, we address efficiency by helping students learn how to obtain the right answer as quickly as possible as is appropriate for the concept. After correctness and speed, we help students improve flexibility so that they understand how to apply their learning to new situations and in different ways. Flexibility means adaptability. We want students to be able to revise and adjust their thinking in ways that maintain and increase accuracy and efficiency. Finally, we must support students as they build their fluency with mathematics. Fluency relates to the confidence and ease with which students work with mathematics concepts.

Accuracy, efficiency, flexibility, and fluency are goals built upon one another within each math concept. Accuracy comes first because there is no point in having any of the other goals without exactness. We do not want students to increase speed if they are getting the answers wrong. Likewise, there is little need to be flexible and fluent with incorrect answers. Educators should focus on accuracy first, even when it takes a long time for students to come up with the correct answer. As soon as students have established accuracy, we need to move right on to focus on efficiency. Our goal is to help students learn how to maintain accuracy while increasing speed. As soon as the students quickly obtain the right answers, we want to focus on flexibility by providing new situations that require changes in their thinking. The fluency aspect of mathematics can be addressed during any of the three previous goals. We can build students' fluency while increasing accuracy, efficiency, and flexibility.

Think about this situation. A young student works on this addition problem $2 + 5 = ?$ In the beginning, the student may use a *count all* strategy. He counts two cubes and five cubes. If he comes up with seven cubes, he has accuracy. However, his strategy was not so efficient because it took him a long time to count all the cubes. To help him improve his speed, we can teach him to *count on*. If he thinks about the first addend and counts on the second addend, he will increase speed. After he can do this successfully, we can teach him to improve flexibility by *counting on starting with the larger addend*. He thinks about five and counts on two more. Because he is able to adapt his thinking to include the commutative property notion that $2 + 5$ is the same as $5 + 2$, he has achieved flexibility and also increased his speed even more because *counting on* from five is faster than *counting on* from two when the sum is seven. Given these and other opportunities to successfully utilize his newly acquired skills, the student increases fluency.

Timing is very important. Sometimes well-meaning math educators jump too fast to efficiency and flexibility. They impose all of these rules and procedures with intentions of helping students see efficient and flexible ways to solve problems. However, teaching rules and procedures is not the same as teaching concepts. Understanding what to do is not enough. We must teach students why a procedure works so that they better understand the concept. If we only teach rules and procedures, struggling students typically achieve random accuracy (if they remember the rules and procedures) and rarely achieve flexibility or fluency because they do not truly own the knowledge. Math educators need to teach concepts to help students build foundations so that they can understand the math. When concepts are ignored and the focus is solely on rules and procedures, struggling students often develop misconceptions and learning gaps. However, if students understand concepts, it is appropriate to help them increase speed and flexibility by teaching rules and procedures, but we must include reasons and connections to help students make this math meaningful.

Students may return to inefficient strategies when faced with challenging problems. They do so because they want to achieve accuracy. These students benefit from the teacher providing intervention that targets their exact academic needs while challenging them to increase accuracy, efficiency, flexibility, and fluency. While it makes sense to focus first on accuracy then move to efficiency, flexibility, and fluency, the goals are not isolated. We often move back and forth, teaching all four goals at the same time to help students understand the hows and whys of mathematics.

Focus on Number Sense and Computation

Typically, the problems that struggling students have in mathematics relate to number sense and computation. Number sense involves an expansive and inclusive understanding of numbers and operations that allows a person to make sound judgments and utilize practical and effective math strategies (McIntosh, Reys, & Reys 1993). Number sense is not something that a person has or does not have. Number sense can be taught through a concept-based approach. It is important to examine all of the aspects of number sense to identify which of these concepts our struggling students are missing. General intervention is not the best way to build students' number sense. We must build students' number sense through targeted intervention aimed at the students' exact academic needs.

The *Focal Points* in mathematics for the elementary grades highlight numbers and operations as critically important and consistent areas of concentration (NCTM 2006). All of the concepts in this book are interrelated components of number sense and operations. This focus on specific concepts does not imply that we need to teach math concepts in isolation. Productive mathematics instruction is inclusive, interconnected, and rich. A focus on number sense and computation concepts in math intervention helps us identify and address gaps and misconceptions that students have in their understanding of mathematics. Intervention does not replace math instruction. Intervention is in addition to good mathematics instruction.

What Makes Math Intervention Successful?

Students who struggle in mathematics need explicit instruction. Explicit instruction is provided when teachers give students ongoing feedback, clear models, a variety of examples, time for practice, and opportunities to talk through the hows and whys of the math situation (National Mathematics Advisory Panel 2008). Successful math intervention is precise and focused on students' immediate learning needs.

Diagnosing and addressing students' needs are part of Response-to-Intervention (RTI). Although the use of RTI is common practice in language arts, RTI is completely applicable to mathematics, as well. RTI promotes the identification and diagnosis of potential learning problems by analyzing students' responses to instruction and providing corrective feedback. With RTI students are offered specific interventions and progress is frequently monitored. Tier 2 (small group) and Tier 3 (individual) are excellent group structures for providing the specific interventions presented in this book. Successful mathematics intervention applies the structure of RTI—identify, diagnose, and address students' learning weaknesses early and precisely.

There are several other key components that make math intervention successful. The group structure, quantity and quality of the tasks, use of new strategies, educator's expertise, incorporation of problem solving, and student's level of motivation are some of the factors that influence math intervention success.

Productive math intervention is rarely a whole class endeavor. Math intervention is most effective in small group structures. Students are more likely to reveal gaps in understanding and misconceptions when they are in small groups. Teachers are more likely to find these gaps and misunderstandings when they are working with a small group. Furthermore, the small group structure allows teachers to target the instruction more effectively because they can constantly adjust the instruction. Teachers can tailor the interaction and math discourse to help struggling students find success. Small group instruction provides opportunities for strategic mathematics intervention.

When possible, one-on-one instruction has powerful potential. The teacher can concentrate completely on one student's academic needs. For some of us, one-on-one instruction is only possible in small chunks of time because we have an entire class to teach. For others such as special educators, tutors, paraeducators, assistants, parents, and volunteers, one-on-one instruction is more common. The benefits of one-on-one instruction can be great because this structure provides opportunities for intensive intervention. Students are typically highly engaged and they gain confidence in their abilities.

More is better is not necessarily true when it comes to math intervention. It really depends on what you mean by *more*. A packet of 100 math problems does not make a better mathematics student. More problems do not necessarily help students, especially if students are left alone to solve page after page of problems. In these cases, all the students typically learn is that they are not good at solving math problems—worse yet, that they hate math. To make the *more is better* idea beneficial for struggling students, consider more time, more instruction, more strategies, and more opportunities for success. These are the ingredients needed for successful math intervention.

Drill does kill. It kills the spirit of learning (Taylor-Cox & Kelly 2008). Too much drill is boring, isolated, and unproductive. Struggling students do not need stacks of flash cards. They need motivating math games that provide them with multiple opportunities for learning. Struggling students must have experiences that help them understand the math, not empty, meaningless experiences. They need context and connections so they learn to own the mathematics.

Certainly, struggling students do not need more of the same. Obviously, the original instruction did not work; otherwise they would not be struggling. Struggling students need new strategies and experiences beyond the textbook. We must offer our struggling students targeted instruction aimed at fortifying specific math concepts for students. Struggling students benefit from new and creative ways in which to learn the information.

An educator's expertise plays an important role in successful math intervention for struggling students. One of the recommendations made on the recent report given by the National Mathematics Advisory Panel involves the expertise of mathematics educators. The panel recommends that "teachers must know in detail the mathematical content they are responsible for teaching and its connections to other important mathematics topics, both prior to and beyond the level they are assigned to teach (2008 p. 37). Educators must understand the concepts, know which concepts the student needs to work on, and know how to assess the level of understanding the student currently has. Good educators strive to increase their levels of expertise. They gain knowledge through their classroom experiences, pedagogical discourse with other educators, and using a variety of resources. Those providing math intervention can be a veteran or new classroom teacher, special educator, ELL teacher, paraeducator, instructional assistant, teacher aide, volunteer, tutor, and/or parent. The point is for educators to have the expertise and resources necessary to provide successful math intervention for students.

Successful math intervention includes problem solving that is interwoven, not a separate component. The goals are to enable students to build understanding, apply and adapt high-quality strategies, and to evaluate and reflect as they solve math problems (NCTM 2000). When students are engaged in meaningful mathematics, they become better problem solvers. Problem solving is embedded within a math situation when students use various strategies, make decisions, and evaluate their thinking.

The level of motivation and commitment that a student has influences the effectiveness of math intervention. Games motivate and empower students. Games open up multiple opportunities for meaningful mathematics, and they are fun. If given a choice, most students would choose to play a game rather than solve math problems on a page in a textbook. Yet the game may actually teach them more about math than they even realize. That's the beauty of games! They are so entertaining that the learning doesn't seem like a chore. This is not to imply that the math will always be easy. Challenges and increased difficulty are important parts of learning mathematics. Math games can be challenging and enjoyable because games serve as a productive way to hook students into successfully engaging in mathematics. The games in this book provide opportunities for students to learn new concepts, practice skills, and apply math thinking to new situations.

How to Use Formative Assessment

Using formative assessment allows us to know what to teach. Formative assessment provides information about what a student currently knows. We can use the formative assessment data to focus on the explicit academic needs of the students.

"Teachers' regular use of formative assessment improves their students' learning, especially if teachers have additional guidance on using the assessment to design and individualize instruction" (National Mathematics Advisory Panel 2008 p. xxiii). Formative assessment comes in a variety of forms. Sometimes it is a performance task. Other times it is a problem or question. Using a formative assessment that is not too time-consuming allows us to devote as much time as possible to the instruction. Interestingly, using formative assessments can actually save time because the information helps the teacher target the instruction. Each of the concepts presented in this book include a formative assessment that will allow the teacher to uncover and focus on the academic needs of the students.

Monitor Progress

Monitoring students' progress is critically important. As educators we must constantly analyze the progress of students. To accomplish this we need to ask questions:

- Is the student making progress?
- What does the student need to learn next?
- How solid is the student's understanding?
- Does the student need more work with a specific concept?
- Is the student having difficulty maintaining and utilizing specific concepts?
- What misconceptions does the student have?
- Where are the learning gaps?
- Is the student's knowledge incomplete? If so, what is missing?

The answers to these questions help us monitor the progress of students. As we monitor progress, we make decisions about how to target the instruction to meet the exact academic needs of the students.

Struggling students need to understand their progress. As educators monitor the progress of students, they should share it with the students. One way to do so is to regularly meet with students and share evidence of the students' gains in knowledge. The evidence may be formative assessment data and/or the actual preassessments and postassessments. The evidence may be a copy of the game board with notes about what a student did to solve problems or a paper that shows how a student used specific concepts and strategies. Many educators use these and other types of evidence to evaluate a student's performance. The key is to provide constant feedback by sharing this information with the student.

Additional ways to help students monitor their own progress include recording the things they know about the concept, have done with the concept, and goals they have related to the concept *(see page 14 for monitoring tool)*. It is also important for students to evaluate the degree of knowledge of any given concept. Using personal scales to report levels of knowledge is beneficial. To do so, the student places X on a scale (the scale on the monitoring tool is 0-10, but any scale is fine) and dates the X underneath. Later, the student adds other Xs and dates to the scale to show progress. Scales offer a productive way to help students focus on their progress.

A former student, Tony, started the school year struggling in mathematics. He had many misconceptions and learning gaps. We worked on building his understanding of specific math concepts by implementing targeted instruction and playing math games to build conceptual knowledge. Evidence of Tony's progress was archived in an accordion folder. The preassessments were stapled to the postassessments to highlight progress. One day Tony was proudly going through the folder. I took the opportunity to join him and we celebrated his progress. After looking through the papers, Tony announced, "I really learned a lot. I can't believe it all fits in my brain!"

Positive ways to share progress with students also include highlighting what they have learned and noting what they need to work on next. The educator may say to the student, "You really improved your accuracy and speed with doubles. The next concept we will focus on is near doubles." The idea is to celebrate what the student has learned and set the next goals.

Reteach

Sometimes students do not learn what they need to learn during regular classroom instruction or intervention. In these cases, we must focus on reteaching. Effective reteaching is not teaching the exact same thing in the exact same way again and again. If students are not learning, we identify the concept and present it in a different way. It may be necessary to move back to a previous concept to build foundations. Or, we may need to reveal and address a misconception. Reteaching should be targeted to meet the explicit academic needs of the students. If a student does not understand fact families, telling the student, in a louder voice, what fact families are is not effective reteaching! Instead, we must try new ways to help the student visualize and model fact families. It may also be that the student needs more work with addition and subtraction concepts. The point is to make the reteaching match the immediate needs of the student.

How to Use This Book

We need to know and understand each concept so that we can identify and focus on specific instruction. This book is designed to give math educators the tools to provide direct math intervention in the areas of number sense and computation.

Students who struggle in mathematics deserve personalized math intervention. Knowing which number concepts a student is lacking (whether the student is four years old or seven years old) is critical. It is not one-size-fits-all math intervention. The point is not to focus so much on the grade level; instead we should focus on the specific concepts that the student needs to learn.

Each math concept presented in this book includes a brief description to provide the educator with background information aimed at increasing expertise. In essence, all of these concepts are connected. Therefore, many of the games include multiple concepts. However, each concept is highlighted independently to enable the educator to offer specific math intervention for students.

A formative assessment is provided for each concept to help the educator find out if the student has gaps in understanding the concept or if the student has a solid understanding of the concept. Sometimes exact quantities and specific numbers are provided. Other times choices are provided. The intent is for the formative assessments to serve as tools for the math intervention.

Successful strategies are included with each number concept for the purpose of offering helpful hints and directions. A math word box accompanies each concept to provide potential math words that should be modeled by the educator and used by the students. These are not vocabulary words to be taught in isolation. Instead, these are conceptually rich math words that should be used in context.

To further promote math discourse and math thinking, sample questions are included. Each question represents a different *level of cognitive demand*. Using Bloom's (1956) taxonomy for structure, a sample question is listed for recall, comprehension, application, analysis, evaluation, and synthesis. The simplest question (recall) is listed first, followed by increasingly advanced levels of questions through the highest level of cognitive demand needed (synthesis). *Note: While Bloom's (1956) original work suggested evaluation as the highest level, my experience with using the levels of cognitive demand suggests synthesizing is more difficult than evaluating. In addition, some questions could represent more than one level of cognitive demand.* The educator should decide which level of questioning best suits the student's needs at the time. Students should be encouraged to ask questions, too.

To motivate students and activate learning, a math game that directly connects to the specific concept is offered. During the game, the educator should encourage students to use math words, ask questions, and explain and justify their thinking. Each game lists necessary materials and step-by-step directions. Each game is intended to be a springboard for active learning. The educator may modify instruction based on how the student responds to various components of the game.

Each game includes potential content differentiation. If the student has too much difficulty with the concept, the educator is invited to differentiate the content of the game for the student by *moving back* to an easier version of the game. If the student understands the concept, the educator is invited to differentiate the content of the game for the student by *moving ahead* to a more complex version.

This math intervention book is a tool for educators. The book includes early number concepts through addition and subtraction designed for pre-kindergarten through second grade students. The companion book *Math Intervention: Building Number Power with Formative Assessments, Differentiation, and Games (Grades 3-5)* includes addition and subtraction concepts through multifaceted number concepts. It is necessary to include addition and subtraction concepts in both books because these ideas are critical to both grade level bands. There may even be situations where both books are needed. Some students may need to begin with the early number concepts. Other students may start with addition and subtraction concepts. The point is to use the formative assessments to find out exactly what the students need...and let the games begin!

_____'s Math Progress

Date/s: _____

Concept: _____

Things I know about this concept:

_____

_____

Things I have done with this concept:

_____

_____

Goals:

_____

_____

_____

How much I know about this concept;
(place X on the scale with date underneath to show level of knowledge; 10 = complete knowledge)

●————————————————————————————●

0   10

CHAPTER

Early Number Concepts

One-to-one Correspondence

Rote Counting

Rational Counting

Keeping Track

Cardinality

Conservation of Number

Subitizing

What is One-to-one Correspondence?

Even though the term "one-to-one correspondence" has been used for over 100 years (OED2), it is often a misunderstood concept. Some educators think that one-to-one correspondence involves successful counting of objects. While a student does need one-to-one correspondence to successfully count objects, the reality is that one-to-one correspondence comes before counting. It involves matching one member of a set with a member of a different set. If a student can place one bowl on each placemat or give one crayon to each student, this student has one-to-one correspondence. We may find that the student has one-to-one correspondence up to a certain quantity. Perhaps the student can successfully match objects when the quantity is five or less. In this case, we say the student has one-to-one correspondence through five. Our focus of instruction for this student would be on quantities greater than five.

 ## Formative Assessment

To find out if a student has one-to-one correspondence, see if the student can successfully match objects. Place five sheets of paper on the table. Give the student five counters. Ask the student to place one counter on each sheet of paper. If the student is successful, increase the quantity of paper and counters. To find out if the student truly owns this concept, provide one more counter than sheets of paper to see what the student does with the situation. Students who own one-to-one correspondence will indicate that they need another sheet of paper to complete the task.

Successful Strategies

Use a variety of objects that are meaningful to students. Begin with smaller quantities and increase the quantities as students gain accuracy and efficiency. To build flexibility and fluency begin with objects that are all the same color, size, and shape. Then move to objects that have different attributes. Encourage students to distribute supplies and materials to each other to provide additional opportunities for one-to-one correspondence.

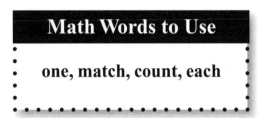

Math Words to Use

one, match, count, each

Questions at Different *Levels of Cognitive Demand*

EASY

Recall: *How did you match one with each?*

Comprehension: *Do all of the objects have a match?*

Application: *How did you use matching?*

Analysis: *How do you know that all of the objects have a match?*

Evaluation: *How would you help someone who could not match each?*

COMPLEX   **Synthesis:** *Is there another way to match each?*

Sleepy Bears

A game designed to build the concept of one-to-one correspondence

Materials:

PLAYERS: 2

- Sleepy Bears Game Board
- Ten bear counters
- Cup

Directions:

1. Place ten bear counters in a cup.
2. Spill the bears out on the table. Notice the "sleepy" bears that are face down.
3. Place each "sleepy" bear in a cave. One bear per cave.
4. Place the rest of the bears back in the cup.
5. Players take turns spilling the rest of the bears and placing the "sleepy" bears in caves.
6. The game is complete when all the bears are placed in separate caves.

Content Differentiation:

Moving Back: Fold the game board in half to play this game using a quantity of five bears.

Moving Ahead: Use two game boards to work with a quantity of twenty bears.

Sleepy Bears Game Board

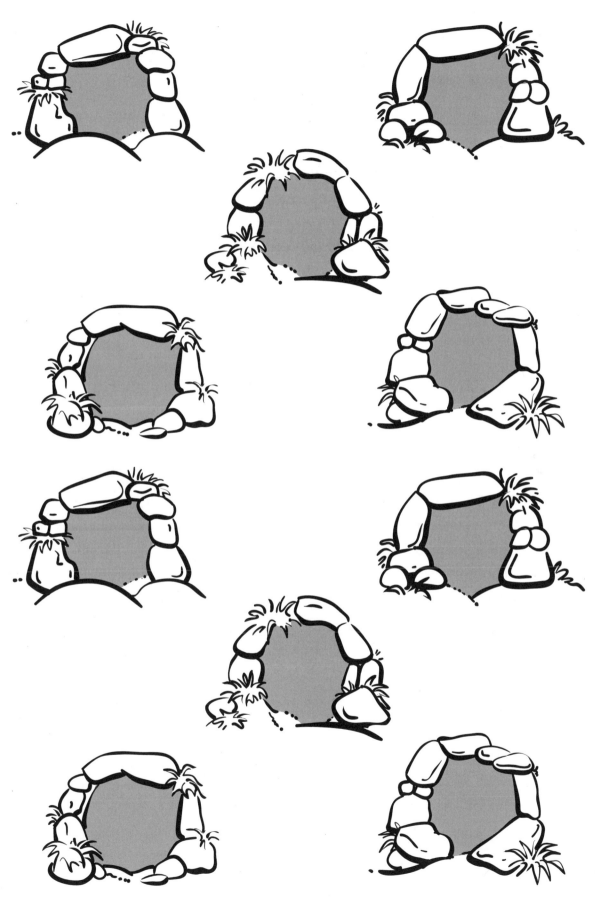

What is Rote Counting?

Rote counting involves saying numbers in order. While rote counting in and of itself offers little context, it is an important step toward understanding contextual counting. The easiest form of rote counting is counting by ones in increasing order. More complex rote counting involves counting in decreasing order because it requires the student to learn a new sequence. To raise the complexity level higher, rote counting can involve skip counting by a constant amount (tens, fives, twos). Rote counting is further complicated when skip counting begins with a number other than zero.

Formative Assessment

To find out how comfortable a student is with rote counting, ask the student to count out loud. Record any omitted or repeated numbers. Notice the student's level of comfort. Does the student use a lot of think time to retrieve the numbers? Or does the student appear comfortable with rote counting? If the student can accurately count by ones in ascending order, try counting by ones in descending order. If appropriate, check to see if the student can count by tens, fives, and twos. Check skip counting starting with zero before checking skip counting starting with one or another number.

Successful Strategies

Learning how to rote count requires hearing others rote counting. Educators need to model rote counting as often as possible. Use different voice levels and/or voice pitches to highlight the counting sequence. "Math-er-cise" with your students by adding hand motions or physical exercise to encourage your kinesthetic learners.

> ### Math Words to Use
>
> **count, before, after, next, between, one, two, three, four, five, six, seven, eight, nine, ten**

 ## Questions at Different *Levels of Cognitive Demand*

EASY

Recall: *What number comes next?*

Comprehension: *Which examples could you give of numbers that come before?*

Application: *How could you show the counting?*

Analysis: *What kind of counting is this?*

Evaluation: *How do you know that a number is left out?*

COMPLEX **Synthesis:** *What if we started with a different number?*

Stomp Your Feet

A game designed to build the concept of rote counting

Materials:

- Timer

PLAYERS: 3 OR MORE

Directions:

1. Set a timer for ten seconds.
2. Altogether players begin stomping their feet as they rote count by ones.
3. When the timer rings, players stop stomping and counting.
4. The first player announces what number comes next.
5. The other players verify the answer.
6. The next player sets the timer for a different amount of time (seconds).
7. Altogether players stomp their feet as they rote count by ones.
8. When the timer rings, the next player announces the number that comes next.
9. Play continues until all players have had several turns.

Content Differentiation:

Moving Back: Begin with fewer seconds to limit the rote counting range. Or repeat the times set on the timer to increase practice with the same rote counting experiences.

Moving Ahead: Try skip counting and other motions. For example, arm stretches by tens, knee bends by fives, toe touches by twos.

What is Rational Counting?

Rational counting is when one-to-one correspondence and rote counting come together in harmony. The student accurately and efficiently counts objects. The counting has context and meaning. Students have ownership of rational counting up to the quantity they name successfully. It is important to know a student's range of rational counting so that the instruction can be targeted for the student.

Formative Assessment

To find out if a student can rational count, provide objects placed in a line. Objects should be the same size, same color, and in the same position. Ask the student to count the objects. Notice if the student touches or slides the objects when counting. If the student accurately counts the objects increase the quantity. If the student does not accurately count the objects, decrease the quantity. Identify the student's rational counting range.

Successful Strategies

Teaching students to slide or drop the objects as they count can be helpful when a student is first learning about rational counting because these motions help the student accurately count. Eventually, however, these strategies erode efficiency because they slow the student down. Likewise, touching objects as one counts can help improve accuracy. But if we do not touch the objects as we count, we increase our speed (efficiency). If a student accurately counts a given quantity of objects, help the student increase speed by encouraging him to count by just looking. If the student inaccurately counts a given quantity of objects, encourage strategies such as sliding, dropping, or touching to help improve accuracy. As the student increases accuracy and efficiency with rational counting, provide situations to work with flexibility and fluency by presenting objects in nonlinear arrangements and/or by using objects with different attributes.

> ### Math Words to Use
>
> **count, many, one, two, three, four, five, six, seven, eight, nine, ten**

 ## Questions at Different *Levels of Cognitive Demand*

EASY

Recall: *How many are there?*

Comprehension: *How did you count?*

Application: *Can you show another way to count?*

Analysis: *Is there a way to show groups as you count?*

Evaluation: *How do you know your way of counting is the fastest way?*

COMPLEX

Synthesis: *What if three were lost?*

Try to Go Home

*A game designed to build the concept
of rational counting*

**PLAYERS:
2**

Materials:

- Try to Go Home Game Board
- Pawn
- Die (1-6 labeled with dots)

Directions:

1. Place one pawn on the center spot of the game board.
2. The first player rolls the die and moves the pawn toward her "Home" spot.
3. The second player rolls the die and moves the same pawn toward her "Home" spot.
4. Play continues until a player reaches her "Home" spot.

Content Differentiation:

Moving Back: Make a new game board with fewer spaces.
Use a die labeled with 1, 1, 2, 2, 3, 3 (instead of 1, 2, 3, 4, 5, 6).

Moving Ahead: Make a new game board with more spaces.
Use two dice and find the sum.

*NOTE: This game appears in Taylor-Cox, J. (2005) **Family Math Night: Math Standards in Action**. Eye on Education.*

Try to Go Home Game Board

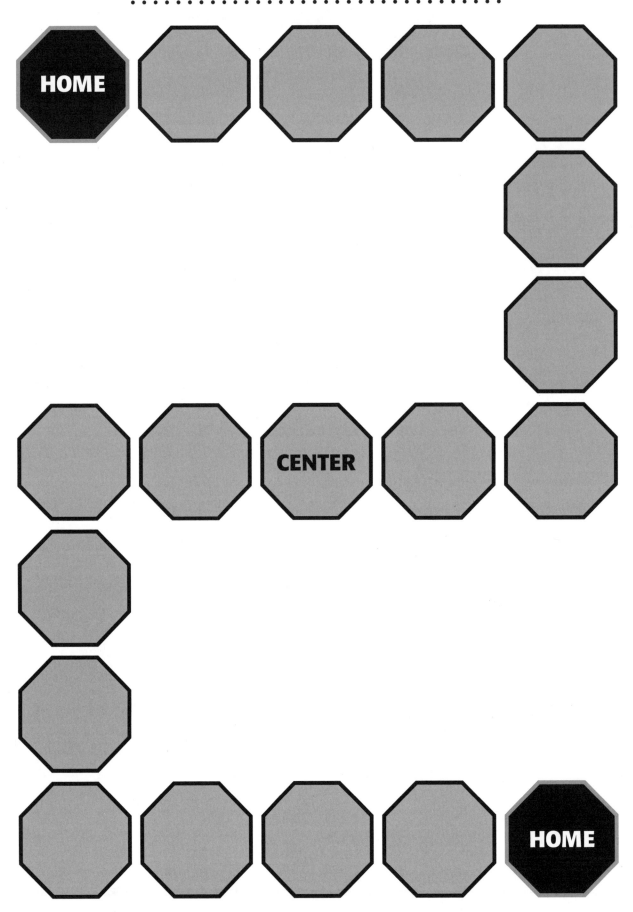

 Math Intervention: Grades PreK-2 • *Jennifer Taylor-Cox* • *© Eye On Education*

What is Keeping Track?

Keeping track involves knowing which objects have been counted and which objects have not yet been counted. Students who do not understand how to keep track often skip objects when counting and/or recount the same objects. To be successful with keeping track, students need to learn how to organize the objects in ways that are helpful. Linear arrangements of objects are often a beneficial initial step.

Formative Assessment

To find out if a student can keep track, provide the student with a pile of ten counters. Ask the student to count the objects. Notice if the student organizes the counters or if the student misses any objects while counting. Notice if the student recounts the same objects. If the student is able to keep track, increase the quantity. If the student is not able to keep track, decrease the quantity. Find out the scope of the student's ability to keep track when objects are presented in non-linear arrangements.

Successful Strategies

Teach students how to line up objects before or during rational counting. As accuracy improves, teach students how to cluster or group objects as they count. Begin with small quantities and move to greater quantities. Increase flexibility and fluency by beginning with objects that are alike and moving to objects that are different.

Math Words to Use

keep track, count, one, two, three, four, five, six, seven, eight, nine, ten

---

Questions at Different *Levels of Cognitive Demand*

EASY

Recall: *How many are there?*

Comprehension: *Were all of the objects counted?*

Application: *What kind of counting did you use?*

Analysis: *How did you keep track? Compare your way to another person's way.*

Evaluation: *Is your way of keeping track a good way? Why?*

Synthesis: *Is there another way to keep track when counting?*

COMPLEX

On the Right Track

*A game designed to build the concept
of keeping track*

Materials:

**PLAYERS:
2**

- ■ 17 small bi-colored counters or
  lima beans painted red on one side
- ■ Cup
- ■ On the Right Track Game Board

Directions:

1. Place the counters/beans in the cup.
2. Spill counters/beans in the center of the track.
3. All of the counters/beans that land with red facing up will be counted by placing each in a space on the railroad track.
4. Players count as they place the counters/beans in the spaces of the track.
5. Replace all counters/beans in the cup.
6. Players take turns spilling and counting the counters/beans.

Content Differentiation:

Moving Back: Use fewer counters/beans.

Moving Ahead: Use more counters/beans by placing two in each space of the railroad track.

On the Right Track Game Board

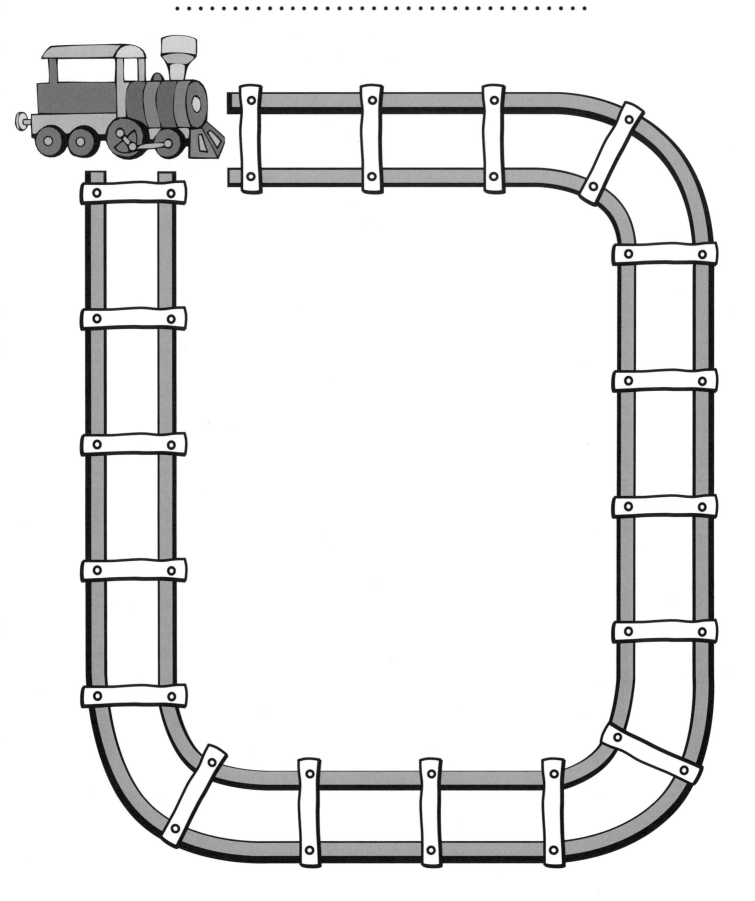

Math Intervention: Grades PreK-2 • Jennifer Taylor-Cox • © Eye On Education

Cardinality

What is Cardinality?

Cardinality involves knowing that the last number named when counting a group of objects is the quantity of the set. This last number named also represents the last object (NCTM, 2000), which makes cardinality a bit confusing for some students. If a student who is asked "how many?" recounts rather than stating the quantity of the set, he does not have cardinality. Students who have cardinality can tell how many after counting. They know that the last number names the last object and the total of the set. Students may have cardinality of some numbers but not other numbers. There may also be levels of cardinality ranging from no understanding to complete understanding (Bermejo, 1996).

Formative Assessment

To find out if a student has cardinality of a number, give the student some counters and ask "How many?" If the student states the quantity, he has cardinality of that number. If the student recounts the objects, he does not have cardinality of the number. Adjust quantity of objects accordingly.

Successful Strategies

Begin with smaller quantities. When counting objects together, the educator should model the language, for example, "One, two, three, four, five. There are five shells." Initially, use counters that are the same size, same shape, and placed in the same position.

As the student successfully shows her skill with cardinality, increase the level of difficulty by using objects that are different colors, shapes, and in different positions.

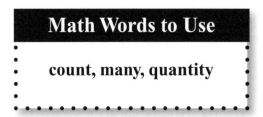

Math Words to Use

count, many, quantity

Questions at Different *Levels of Cognitive Demand*

EASY

Recall: *How many?*

Comprehension: *How would you explain how you know the quantity?*

Application: *Can you show the quantity another way?*

Analysis: *How many parts are in the quantity?*

Evaluation: *How would you help someone who could not tell how many?*

COMPLEX

Synthesis: *How could you invent a new way of counting?*

A Rainbow of Beads

A game designed to build the concept of cardinality

Materials:

- String (or pipe cleaners) for each player
- Bowl of beads (six colors)
- Die (labeled 1-6 with numerals)
- Color cube (one color on each side, use colors that correspond to the bead colors)
- Sheet of paper

PLAYERS: 2 OR MORE

Directions:

1. The first player rolls the color cube and the die, then covers these with a sheet of paper.
2. The player strings and counts aloud the corresponding number of colored beads.
3. The other players ask, "How many _____ beads?"
   (color rolled)
4. The first player tells how many beads. If the player has to recount the beads to name the quantity or peek under the sheet of paper, the beads are removed and placed back into the bowl. If not, the player keeps the beads on the string.
5. Players take turns rolling the cubes, stringing the beads, and telling how many beads of each color.
6. If a player rolls a color that he already has on the string, he misses the turn.
7. The first player to roll, string, and state the quantities of all six colors is the winner.

Content Differentiation:

Moving Back: Use dice with dots or dice labeled 1, 1, 2, 2, 3, 3.

Moving Ahead: Use dice labeled with larger numerals or use two dice and find the sum.

CONCEPT:
Conservation of Number

 ## What is Conservation of Number?

Conservation of number involves understanding that a quantity remains the same even when it is arranged differently (Piaget, 1964, 1970). Students who do not yet conserve often believe that a quantity is greater when the objects are arranged further apart. Likewise, these students believe that the quantity is smaller when the objects are arranged closer together. Students who do not yet have conservation of number can be distracted by the arrangement of objects, even if they actually observe the change in arrangement. Trying to convince a student that the arrangement does not impact the quantity is not the best approach. Because conservation is a cognitive process, we need to give students more meaningful experiences to help them achieve conservation of number.

 ## Formative Assessment

To find out if a student has conservation of number, give the student five cubes and give yourself five cubes. Ask the student how many she has and how many you have. After the student responds (saying that each has five), arrange the student's five cubes in a line with space between the cubes. Arrange your five cubes in a cluster. Ask the student, "Who has more?" If the student does not have conservation of number, she will announce that she now has more. If the student does have conservation of number, she will respond that both sets are still the same, or five, or equal. The student who has complete ownership of conservation of number may even find it amusing that you are asking "Who has more?" because they know that the quantities have not changed. Repeat the task with larger quantities to determine the range of the student's ability to conserve number.

Successful Strategies

Use objects and situations that are meaningful to students. Students are often more "tuned in" when the situation is interesting to them. Even very young students notice when they believe that someone has "more" of something that is desirable (e.g., food, toys). Use motivating situations to teach students how to conserve number.

> ### Math Words to Use
>
> more, less, equal, same,
> quantity, amount, set

Questions at Different *Levels of Cognitive Demand*

EASY

Recall: *How many do you have?*

Comprehension: *How would you explain the change in arrangement?*

Application: *Do you still have the same amount?*

Analysis: *How would you compare my set to your set?*

Evaluation: *Why are the amounts still equal?*

COMPLEX **Synthesis:** *What if we each had half as many?*

Same or Not the Same

*A game designed to build the concept
of conservation of number*

**PLAYERS:
2**

Materials:

- Counters
- Die (labeled 1-6 with numerals)
- Trays
- Same or Not the Same Spinner
- Paper clip and pencil
- Folder

Directions:

1. The first player rolls the die and displays that number of counters on his tray.
2. The second player secretly spins the spinner behind the folder.
3. If the spinner lands on "Same" the player displays the same number of counters on his tray. *NOTE: The counters do not have to be arranged in the same way.*
4. If the spinner lands on "Not the Same," the player displays a different amount of counters on his tray.
5. The first player looks at the two quantities of counters and announces "Same" or "Not the Same."
6. The second player reveals the spinner by lifting the folder up.
7. Players switch roles and continue the game.

Content Differentiation:

Moving Back: Try using similar arrangements when the spinner lands on "Same."

Moving Ahead: Increase quantities by using polyhedral dice.

Same or Not the Same Spinner

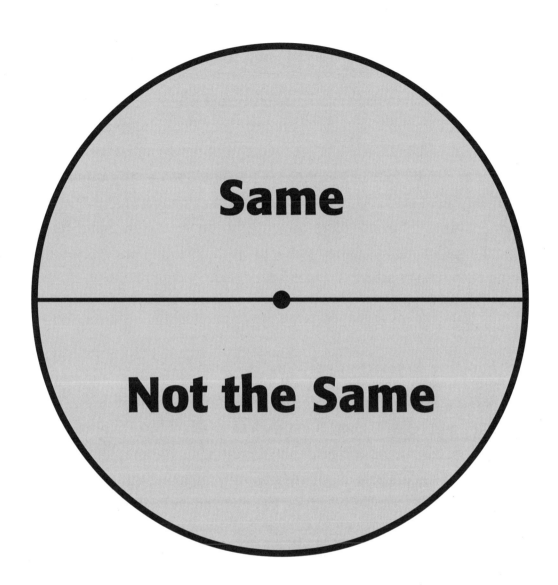

To use the spinner, place a paper clip in the center. Place pencil point through the paper clip at the center of the spinner. Holding the pencil securely with one hand, spin the paper clip with the other hand.

Subitizing

What is Subitizing?

Subitizing involves knowing the quantity instantly without counting. When students subitize, they know how many because the arrangement is familiar and/or friendly. While recent information about the benefits of teaching subitizing has surfaced (Clements, 1999), early use of the term appeared more than fifty years ago (Kauffman, et al., 1949) and early work with the idea dates back nearly 100 years (Freeman, 1912). Teaching students to subitize builds fluency, flexibility, accuracy, and efficiency with numbers. We want students to realize that they can subitize!

Formative Assessment

To find out if a student can subitize, roll a die and ask how many dots are showing. You will know if the student subitizes by how quickly and accurately the student responds. Counting takes time. Subitizing is instant. The dots on a die offer familiar arrangements that can be subitized. Other arrangements can be subitized, as well. If the student can subitize the dots on a die, try dot stickers in rows of three or dot stickers on a ten frame. Determine whether the student can subitize these quantities and arrangements by "flashing" the dots for three seconds and asking the student to tell how many. To gain even further insight, ask the student to explain how she knows. Identify which quantities the student can subitize and which quantities the student cannot subitize.

Successful Strategies

Begin with quantities and arrangements that students can successfully subitize. Discuss how the students know the quantity. Add one more dot (or other symbol/object) to a familiar arrangement and examine how the arrangement is now different. If students need to count the dots (or other symbols/objects), do not discourage them from doing so. When they count instead of subitize, it means that they do not yet have confidence in their accuracy. They need more experiences. They need to build familiarity and confidence with the arrangements. Keep in mind that only some quantities can be subitized. Other quantities must be counted or estimated. The rule of thumb is that if you can subitize the quantity (without estimating or counting), you can teach your students to subitize the quantity.

> ### Math Words to Use
>
> **many, subitize,
> arrangement,
> quantity, total**

 ## Questions at Different *Levels of Cognitive Demand*

EASY

Recall: *How many dots do you see?*

Comprehension: *How do you know?*

Application: *Is the arrangement familiar?*

Analysis: *Which parts of the quantity do you know without counting?*

Evaluation: *Is there a faster way to tell how many?*

Synthesis: *Can you think of a new way to arrange ten so it can be subitized?*

COMPLEX

Realize You Can Subitize!

*A game designed to build the concept
of subitizing*

Materials:

- **Paper dominoes (cut each domino in
  half to show separate regions)**

**PLAYERS:
2 OR MORE**

Directions:

1. **Place all domino regions face down.**
2. **The first player turns over a domino region and begins
   whispering, "Subitize."**
3. **If the second player can name how many dots are shown on
   the domino region before the other player finishes saying,
   "Subitize," he claims the domino region.**
4. **If the player does not name the number of dots before the
   other player finishes saying "Subitize," the domino region
   is placed face down with the remaining domino regions.**
5. **Players take turns.**
6. **The game is over when there are zero domino regions left
   face down.**
7. **The winner is the player who claimed the most domino
   regions.**

Content Differentiation:

Moving Back: **Remove some of the difficult domino regions.
Or say "Subitize" very slowly.**

Moving Ahead: **Use dominoes that are not cut in half and try
subitizing both regions together. Or try double-nine domino
regions.**

Dominoes Double-Six Set
· ·

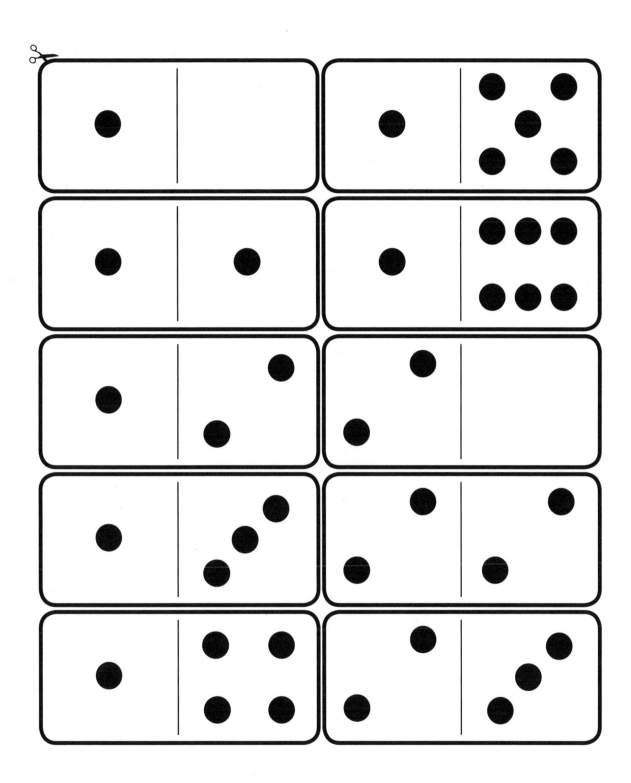

Dominoes Double-Six Set

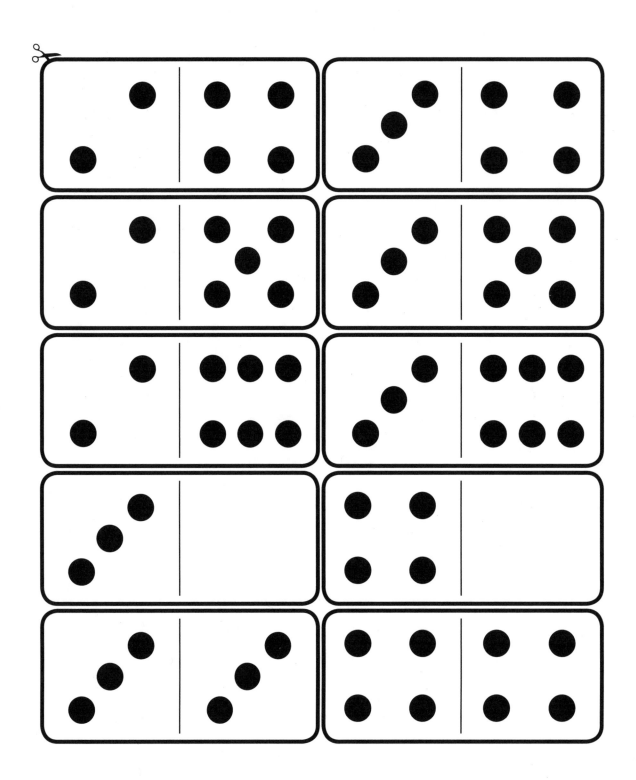

Dominoes Double-Six Set

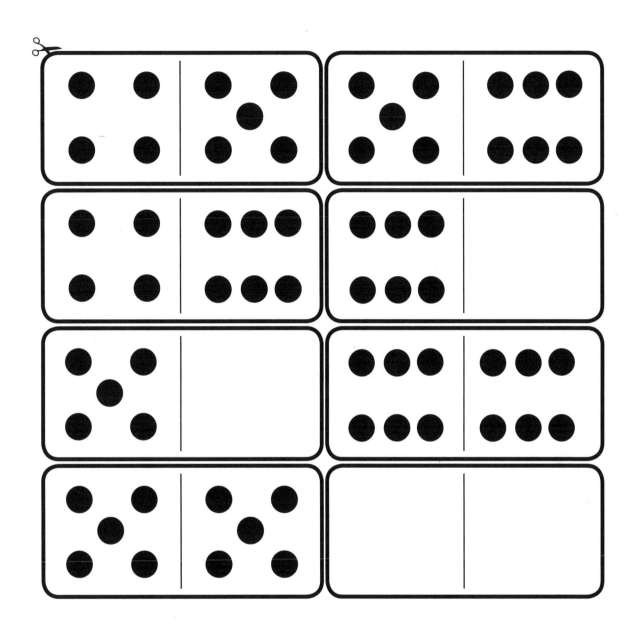

CHAPTER

Numbers and Number Relationships Concepts

Representing Numbers

More and Less

Equal and Unequal

Composing and Decomposing Numbers

Understanding Ten

Ordinal Numbers

Even and Odd

Basic Place Value

Basic Fractions

Estimation

Representing Numbers

What Does it Mean to Represent Numbers?

Representing numbers involves showing numbers in a variety of ways. Numbers can be represented with digits, words, symbols, pictures, and objects. To represent the number seven, for example, a student can write 7 or seven. Students can show seven tally marks, seven stars, seven smiley faces, or draw a picture of a birthday cake with seven candles on it. In addition to digits, words, symbols, and pictures, students can also show objects in different arrangements. Students may represent seven with seven chips on a ten frame, seven linking cubes in a tower, or seven fingers. More complex ways to represent numbers may involve showing the different ways to make seven. For example, six blue bears and one red bear or four fingers on one hand and three fingers on the other hand. The goal is for students to be able to represent numbers in a variety of ways which allows the student to think flexibly about the number.

Formative Assessment

To find out if a student can represent numbers, provide materials such as paper, writing tools, counters, base ten blocks, ten frames, and cubes. Ask the student to show nine using any of the materials. If the student successfully shows nine, ask him to show nine a different way. Find out how many different ways the student can show nine. If the student does not successfully show nine, try a smaller number. If the student successfully shows nine in a variety of ways, try larger numbers.

Successful Strategies

Representing numbers in different ways does not mean just using different counters. Six green bears lined up in a row are not much different from six yellow chips lined up in a row. A student may think it is a different way because she used different objects. However, this is not the flexibility with representing numbers that we are after. However, if the student shows three blue bears and three red bears to represent six then shows six chips arranged in groups of two, six has been represented two different ways. If you are unclear about how the student is representing a number, ask the student to explain the representation. Encourage students to use symbols and pictures that are meaningful to them.

> ### Math Words to Use
>
> **represent, show, number, symbol, numeral, digit**

Questions at Different *Levels of Cognitive Demand*

EASY

Recall: *What number is this?*

Comprehension: *How could you show that number with pictures?*

Application: *Can you show the number another way?*

Analysis: *How does your representation show parts of the number?*

Evaluation: *Is your way to show the number a good way? Why or why not?*

Synthesis: *How could you reorganize your representation?*

COMPLEX

Show Me the Number

*A game designed to build the concept
of representing numbers*

**PLAYERS:
4**

Materials:

- **Decahedron die (labeled 0-9 with numerals)
  or 0-9 spinner** *(see page 62)*
- **Sticky notes and marker**
- **Ten frame and chips**
- **Craft sticks (popsicle sticks)**
- **Counters**

Directions:

1. The first player rolls the die, says the number, and writes the number on a sticky note.
2. The second player shows the number with chips on a ten frame.
3. The third player shows the number with craft sticks arranged as tally marks.
4. The fourth player shows the number with groups of counters (different colors and/or arrangements).
5. Players verify all of the representations.
6. Players switch roles and take turns representing numbers in different ways.
7. If a player rolls a number that was previously represented, the player rolls again.
8. As the game continues, students put the numbers on sticky notes in order.
9. Play continues until all of the numbers on the die have been represented.

Content Differentiation:

Moving Back: Represent numbers in only two ways.

Moving Ahead: Represent 2-digit numbers using a die for tens and a die for ones. Include base ten blocks as another way to represent the numbers.

Ten Frames

CONCEPTS:
More and Less

What are the Concepts of More and Less?

To help students understand numbers and number relationships, they need to identify, create, and compare quantities and values. The concepts of more and less are interconnected and only understood through comparison and/or ordering. Students understand more as the opposite of less. More and less are inter-relational because a quantity is more if it is compared to a quantity that is less. The same quantity could be labeled less/fewer if it is compared to a greater quantity.

Formative Assessment

To find out if a student understands more and less take ten counters from a basket of counters. Ask the student to use the counters in the basket to give himself less/fewer than your set. Notice if the student counts your counters. Watch for any strategies the student may use to compare the quantities. If the student does not show fewer counters, try a smaller quantity. If the student does show a quantity that is less/fewer than yours, write a 2-digit number (such as 25) and ask the student to say a number that is more than the number you wrote. If the student does not accurately identify a greater value, show 25 counters and ask the student to show more using objects. If the attempt is unsuccessful, try smaller numbers.

Successful Strategies

Encourage students to show *more*, *not more*, *a lot more*, and *a little more* in each comparison to a specified amount. Because young students often identify more before they identify less, encourage them to understand less as not more and not equal. Have students show *less*, *not less*, *a lot less*, *a little less* as they compare quantities and ask them to explain how they know. Comparing sets of objects is a powerful way to help students build understanding of more and less. To make the task more complex, give the student more than two quantities/values to compare. Ask the student to order the quantities/values from less (fewer) to more (greater) or vice versa.

> ### Math Words to Use
>
> **more, less, greater, fewer, compare, set, quantity, amount, value**

Questions at Different *Levels of Cognitive Demand*

EASY

Recall: *Which set is more?*

Comprehension: *How do you know it is more?*

Application: *How could you show that this set has more than that set?*

Analysis: *How much less/fewer is this value than that value?*

Evaluation: *Is your way of proving that a set is more or less a good way? Why?*

Synthesis: *How would you order several sets to show greater to fewer amounts?*

COMPLEX

More or Less for Lunch

A game designed to build the concepts of more and less

Materials:

PLAYERS: 2

- More or Less Game Board
- 2 Spinners (labeled 1-8 with numerals)
- 2 paperclips and 2 pencils
- Bowl of 16 lima beans

Directions:

1. Both players spin a spinner at the same time.
2. Players take the corresponding number of lima beans from the bowl.
3. Players decide which set of beans is more/less and place the lima beans on the correct plate (more or less).
4. Players say a comparison expression. For example, "Five is more than three" or "Three is less than five."
5. Players remove the lima beans and place them back in the bowl.
6. Players continue spinning, placing lima beans, and comparing the sets.
7. If both players spin the same number, players must spin again because the lima beans can not be placed. When this happens, the players announce, "No lima beans for lunch!"

Content Differentiation:

Moving Back: Instead of placing lima beans on the plates, line up the lima beans in side-by-side columns to help students see which column of beans is more or less.

Moving Ahead: Change the spinner by writing a zero after each digit, creating quantities of 10-80. Instead of placing lima beans, laminate the game board and have students write the number or make tally marks of each quantity on the plates.

More or Less Game Board

More or Less Spinners

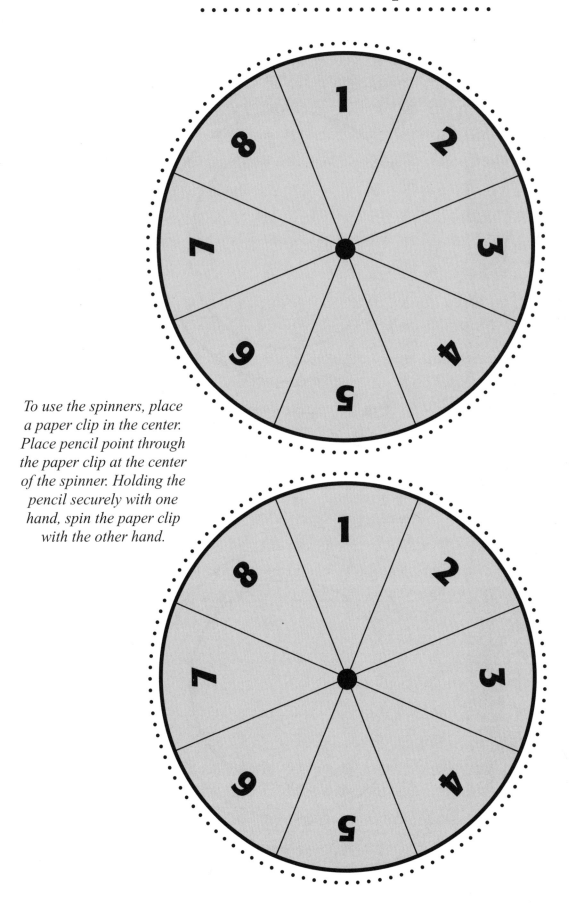

To use the spinners, place a paper clip in the center. Place pencil point through the paper clip at the center of the spinner. Holding the pencil securely with one hand, spin the paper clip with the other hand.

What are Equal and Unequal?

Essentially, equal means the *same as*. When we say, $4 = 3 + 1$, we are saying that four is the *same as* three and one. Specifically, four is the *same value as* three plus one. Even though many students think so, equal does not mean *the answer is coming up next*. It is important that students understand equality via the notion of balance. One side of the equation is the same value as the other side of the equation. It does not matter if the sum is first ($10 = 7 + 3$) or if the addends are first ($7 + 3 = 10$) or if there are multiple addends ($5 + 2 + 3 = 9 + 1$). Equal involves balance. Conversely, unequal means there is no balance because the values are not the same. Understanding what unequal ($\neq$) means helps students understand what equal ($=$) means.

Formative Assessment

To find out if a student understands equal, give the student two trays and a basket of 15 counters. Ask the student to show two equal amounts. Encourage the student to show one amount on one tray and the other amount on the other tray. Even though they do not need to use all the counters to show two equal sets, some students try to do so. That is why it is important to use an odd number of counters. We want to know if the student understands the concept of equal so we give a situation in which the student has to use her knowledge. If the student successfully completes the task, ask if the student knows how to write the equal symbol. Also, check for understanding of unequal by asking the student what changes could be made to the equal sets to make them unequal. If the student does not successfully complete the first part of the task (making two equal sets) use a smaller quantity (seven) to find out if she understands equality with smaller values.

Successful Strategies

Understanding equal requires many experiences with objects in sets. Use the words "equal" and "same as" interchangeably during the student's experiences. Remember to vary the arrangement of objects (lines, piles, stacks, clusters, etc.). As with the concept of equal, understanding unequal requires conservation skills. Students gain knowledge of the concept of unequal as they learn *not equal* and *not the same as*.

> ### Math Words to Use
>
> **equal, unequal, same as, not the same as, balance, value, amount, set**

 ## Questions at Different *Levels of Cognitive Demand*

EASY

Recall: *Are the sets equal?*

Comprehension: *How are the sets balanced?*

Application: *How could you show unequal sets?*

Analysis: *How would you compare this set with several unequal sets?*

Evaluation: *How can you prove that the sets are equal?*

Synthesis: *What kind of changes can you make to show equal/unequal amounts?*

COMPLEX

Princess Ring

*A game designed to build the concepts
of equal and unequal (and more and less)*

Materials:

- Set of double six dominoes *(see page 41)*
- Princess Ring Game Board

**PLAYERS:
2 OR MORE**

Directions:

1. Players place all dominoes face down so that the dots cannot be seen.
2. A player chooses one domino at random to be the Princess and places it in the Princess Ring.
3. The first player chooses a domino and decides if the number of dots is equal to the Princess or unequal to the Princess. If the domino is unequal to the Princess the player decides if it is more or less.
4. The player places the domino on the correct section of the game board.
5. Players take turns choosing dominoes and placing them on the game board. *NOTE: To view all placed dominoes, the game board can be illustrated on large paper.*
6. Players talk about how they know that the domino is equal or unequal to the Princess.
7. Continue playing until all of the dominoes are placed on the board.

Content Differentiation:

Moving Back: Use only the dominoes with smaller values.

Moving Ahead: Discuss "how much more than the princess" or "how many less/fewer than the princess" each unequal domino is.

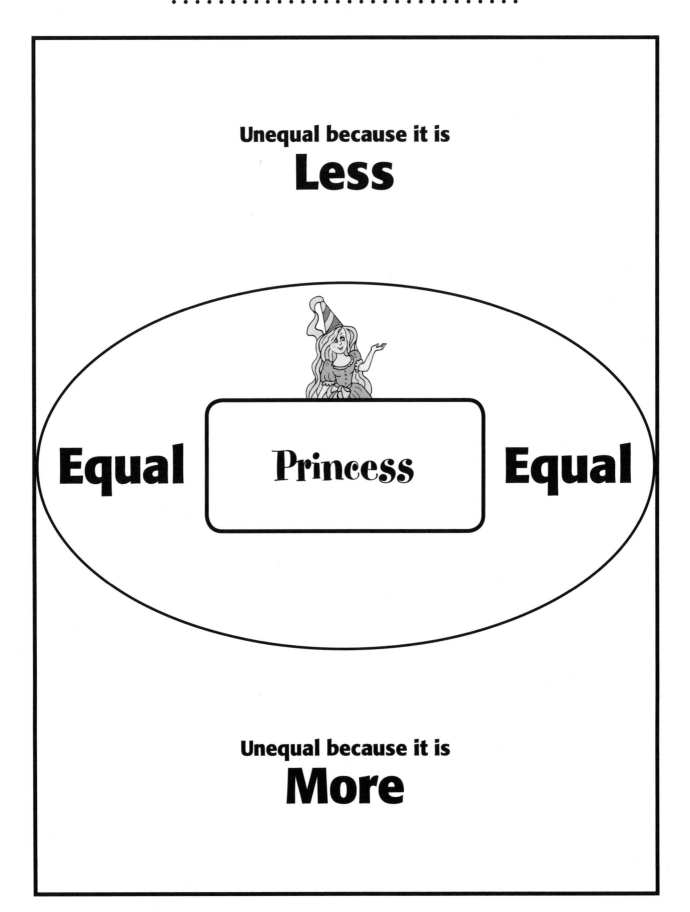

Composing and Decomposing Numbers

What Does it Mean to Compose and Decompose Numbers?

Composing numbers means building numbers up. Decomposing numbers means breaking numbers down. When we make a number, we are composing it. When we take a number apart, we are decomposing it. When students think about how specific numbers are composed and decomposed, it helps them understand numbers, number relationships, and computation.

Formative Assessment

To find out if a student can compose and decompose numbers give the student a tower of 15 connecting cubes of all the same color. Ask the student to show how 15 could be broken down. If the student breaks apart the connecting cube tower, ask him to describe the decomposed parts of 15. If the student is successful, ask him to decompose (break down) 15 a different way. If successful, ask the student to show how 18 could be put together (composed) given a set of 20 or more loose cubes of various colors. Ask the student to show how 18 could be composed another way. If successful, ask the student to try the same tasks with a larger 2-digit number. If unsuccessful, try the tasks with a single-digit number.

Successful Strategies

It is important for students to be able to compose and decompose in various ways. We do not want the student to think that the only way to make 25, for example, is with two tens and five ones. We can also show 25 with one ten and fifteen ones (which is important when adding and subtracting with regrouping). We can also show 25 with five sets of five (which is important when multiplying and dividing). Knowing how to compose and decompose numbers helps students with accuracy, efficiency, flexibility, and fluency.

> **Math Words to Use**
>
> **compose, build up,
> put together, decompose,
> break down, take apart**

Questions at Different *Levels of Cognitive Demand*

EASY

Recall: *What number is it?*

Comprehension: *How would you decompose this number?*

Application: *How could you use cubes to show how to compose this number?*

Analysis: *Is there another way to compose/decompose this number?*

Evaluation: *Why is it important to know different ways to compose/decompose numbers?*

Synthesis: *What if there were three more parts to the number?*

COMPLEX

Four to Score

***A game designed to build the concepts
of composing and decomposing numbers***

**PLAYERS:
2**

Materials:

- **Decahedron die (ten-sided labeled 0-9
  with numerals)** *NOTE: If a decahedron die
  is not available use a 0-9 spinner (see page 62)*
- **Four to Score Game Board**
- **40 translucent chips of one color for each player**

Directions:

1. **The first player rolls the decahedron die and uses chips to
   cover two numbers that show how to decompose the number
   you rolled. For example, if a 5 is rolled, chips could be placed
   on 5 & 0, 4 & 1, or 2 & 3.**
2. **Players take turns rolling the decahedron die and placing
   chips to show ways to decompose the numbers rolled. Chips
   cannot be moved once they are placed on the game board.**
3. **If four chips of the same color are in a row (horizontal) or
   column (vertical), the player scores 1 point. Each chip has the
   potential to be part of several "four in a row/column" points.**
4. **If the number rolled cannot be decomposed by the numbers
   left on the game board, the player misses his turn.**
5. **Continue play until the board is covered. The winner is the
   player with the most points.**

Content Differentiation:

Moving Back: **Leave out the "four in a row/column" aspect of
the game. The winner is the player with the most chips on the
board at the end of the game.**

Moving Ahead: **Try placing more than two chips on the board
to show how to decompose the number (for example, if a 7
is rolled, chips could be placed on 2, 1, and 4). Or include
diagonal four as another way to score a point.**

Optional 0-9 Spinner

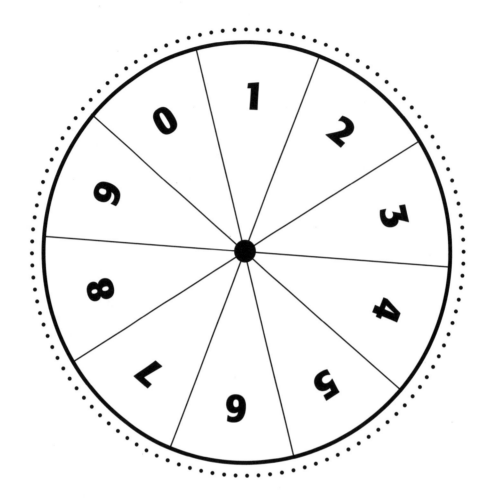

To use the spinner, place a paper clip in the center. Place pencil point through the paper clip at the center of the spinner. Holding the pencil securely with one hand, spin the paper clip with the other hand.

Four to Score Game Board

0	6	1	3	2	4
3	2	0	8	1	5
1	7	3	2	0	6
3	0	5	1	4	2
2	4	3	0	3	1
5	1	2	9	7	0
0	6	1	4	2	3
8	2	0	5	1	4
1	7	4	2	0	6
4	0	3	1	5	0

Understanding Ten

 ## What Does it Mean to Understand Ten?

Ten is the most important number in our base-ten system. Students need to understand the parts of ten. They also need to know how ten relates to other numbers. When students completely understand ten, it means that they own ten. Ownership of ten is one of the best ways to build computational accuracy, efficiency, flexibility, and fluency.

 ## Formative Assessment

To find out if a student understands ten, use a set of ten frame dot cards. Show a ten frame card with nine dots. Ask the student how many dots on the ten frame. If the student responds accurately and quickly, ask her to explain how she knows it is nine. Solid explanations include nine as one less than ten or ten as one more than nine or a row of five and a row of four. Continue showing ten frame cards (with different amounts of dots) in random order. Notice which quantities students can name accurately and efficiently. Use students' explanations to determine the degree to which their understandings are solid.

Successful Strategies

Placing objects on the ten frame helps students visualize ten and how other quantities are related to ten. Some students visualize quantities best when the ten frame is horizontal, while other students visualize quantities best when the ten frame is presented vertically. Focus discussions on how the objects look when the ten frame is full and when it is not full. Also, work with ten objects without the ten frame, allowing students to explore objects in various formations (i.e., groups, piles, stacks, line/s). Extend conversations around ten versus not ten and showing ten in various ways. Encourage students to understand the importance of ten in the base-ten number system.

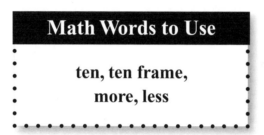

Math Words to Use

**ten, ten frame,
more, less**

 ## Questions at Different *Levels of Cognitive Demand*

EASY **Recall:** *How many dots are on the ten frame?*

Comprehension: *How would you explain how nine looks on the ten frame?*

Application: *How can you show two less than ten?*

Analysis: *Is there another way to show six on the ten frame?*

Evaluation: *Why is the ten frame a good way to show numbers?*

COMPLEX **Synthesis:** *What would a twelve frame look like?*

Match and Claim

A game designed to build the concept of understanding ten

PLAYERS: 2 OR MORE

Materials:

- Set of Ten Frame Dot Cards (copied on cardstock)
- Set of 0-10 Number Cards (copied on cardstock)

Directions:

1. Shuffle the ten frame dot cards and the number cards together.
2. Place cards face down in rows and columns.
3. The first player flips over two cards.
4. If the cards match (number of dots on the ten frame card is the same as the number on the number card), the player claims the cards.
5. If the two cards do not match, the player flips these cards back over and the turn is complete.
6. Players continue taking turns flipping cards trying to obtain a match.
7. The game ends when all the cards are claimed.
8. The winner is the player with the most claimed cards.

Content Differentiation:

Moving Back: If the cards do not match, leave them "face up" rather than flipping them "face down."

Moving Ahead: Use two sets of number and ten frame dot cards and/or double ten frame dot cards and number cards 0-20.

Set of Ten Frame Dot Cards

Number Cards

. . . .

10	4
9	3
8	2
7	1
6	0
5	

CONCEPT:
Ordinal Numbers

 What are Ordinal Numbers?

Ordinal numbers are numbers that designate the place and position in an ordered arrangement. The first position precedes the second position. Placements continue with third, fourth, fifth, sixth, etc. Yes, the life of a math teacher would be easier if the ordinal numbers began with oneth followed by twoth. But, alas, ordinal numbers are not designated in that way. We need to give students many experiences with ordinal number vocabulary and concepts to build accuracy, efficiency, flexibility, and fluency.

 Formative Assessment

To find out if a student knows ordinal numbers, place ten objects in a row and ask the student to describe the ordinal placements. If needed, tell the student that the object before all others is "first." Ask what comes after first and encourage the student to continue naming the objects with ordinal numbers. If successful through tenth, ask the student what comes next. Continue through 21st. Notice accuracy and efficiency. Invite the student to think flexibly and fluently by asking which ordinal number comes after 99th or which number comes before 72nd.

Successful Strategies

Using objects is a beneficial way to help students practice ordinal numbers. When the objects are presented in an organized way (linear arrangement), students are more likely to apply ordinal number knowledge. Students can also line up themselves and name their ordinal number placements to add meaning to the experiences with ordinal numbers. Additionally, the calendar offers an excellent tool for real life connections with ordinal numbers.

Math Words to Use

ordinal numbers, position, order, first, second, third, fourth, fifth, sixth, seventh, eighth, ninth, tenth

Questions at Different *Levels of Cognitive Demand*

EASY

Recall: *What ordinal number comes after third?*

Comprehension: *How does fifth relate to five?*

Application: *How could you use ordinal numbers to describe today's date?*

Analysis: *How could you categorize the first ten ordinal numbers?*

Evaluation: *How would you help someone who could not remember ordinal numbers?*

Synthesis: *What if the ordinal numbers had different names?*

COMPLEX

Elephant Parade

A game designed to build the concept of ordinal numbers

PLAYERS: 2 OR MORE

Materials:

- Elephant Parade Game Board
- Elephant Cards

Directions:

1. Players find the 1st elephant card and place it in the first box on the Elephant Parade game board.
2. Players mix up all other elephant cards and place them "face down" in a pile.
3. Players take turns drawing an elephant card and placing it in the correct place on the game board.
4. If a player places the last elephant card that completes any of the (horizontal) rows on the game board, one point is earned.
5. Play continues until all elephant cards are placed correctly on the game board.
6. The winner is the player with the most points.

Content Differentiation:

Moving Back: Cut apart the rows of the game board and tape the ends together creating one long horizontal line of places to put the elephant cards. Use fewer elephant cards (for example 1st through 10th).

Moving Ahead: Place the tenth elephant card in any box (other than the tenth) on the game board to start the game. If any elephant cards that cannot be played are drawn, the player places the card in the discard pile and loses her turn. Or players can roll a die to determine how many cards to take from the pile on each turn.

Elephant Parade Game Board

· ·

Math Intervention: Grades PreK-2 • *Jennifer Taylor-Cox* • *© Eye On Education*

Elephant Cards

· · · · · · · · · · · · · · · · ·

1st	2nd	3rd	4th	5th	6th	7th
8th	9th	10th	11th	12th	13th	14th
15th	16th	17th	18th	19th	20th	21st
22nd	23rd	24th	25th	26th	27th	28th
29th	30th	31st	32nd	33rd	34th	35th

Even and Odd

 ## What are Even and Odd?

Odd and even are concepts that relate to patterns and number relationships. To help students understand this concept, compare odd and even amounts. Even amounts always have "partners" and odd amounts always have a "left over" or "extra." Students are more likely to see these comparative relationships when they are able to line up objects in pairs.

 ## Formative Assessment

To find out if a student understands even and odd, ask the student if five is an even or an odd number. Ask the student to explain (or demonstrate with objects) how she knows that five is an odd number. If the student is successful, ask the same questions using larger numbers (for example, 12, 29, 37, 46, 80).

Successful Strategies

Encourage students to describe quantities/values as odd or even. Students can "prove" their answers by lining up the objects in pairs. Students can also show odd/even patterns on number lines and hundred charts. When students are ready, encourage them to explore the relationship between and among odd and even numbers (as when two odds are added together, two evens are added together, or when an odd is added to an even).

Math Words to Use

even, odd, pair, partners, same, not the same, equal, unequal

---

 ## Questions at Different *Levels of Cognitive Demand*

EASY

Recall: *Is the number even or odd?*

Comprehension: *How do you know that the number is even/odd?*

Application: *How could you show that the next number is even/odd?*

Analysis: *What happens when you add an even number to an odd number?*

Evaluation: *Will you always get an even number when you add two odd numbers together?*

Synthesis: *How could you design a key/legend for naming odd and even numbers?*

COMPLEX

Shoes

A game designed to build the concepts of even and odd

Materials:

- Shoe Chart
- 20 cubes (same size and color)
- Icosahedron die (20 sides) If an icosahedron die is not available use a 1-20 spinner *(see page 78)*

PLAYERS: 2 OR MORE

Directions:

1. The first player rolls the die to find out how many shoes. The player says, "I rolled ___ shoes."
2. The player counts out cubes to represent the shoes and places the cubes on the Shoe Chart. If all of the cubes (shoes) are paired up, the number is even. If one of the cubes (shoes) is not paired up, the number is odd. The player tells whether the number of shoes is even or odd. *NOTE: Be sure to make pairs as the cubes are placed. If needed, each pair of shoes can be colored differently to help distinguish each pair.*
3. Players take turns rolling the die, counting cubes, and placing the cubes on the Shoe Chart.
4. Players are encouraged to make predictions about odd and even before verifying by placing the cubes on the Shoe Chart.
5. The winner is the first player to roll 20 and correctly describe it as even.

Content Differentiation:

Moving Back: Work with smaller numbers (decahedron die and half of the Shoe Chart).

Moving Ahead: Use two dice (labeled 1-6 with numerals). Add numbers together examining the patterns of adding odd and even numbers.

Shoe Chart

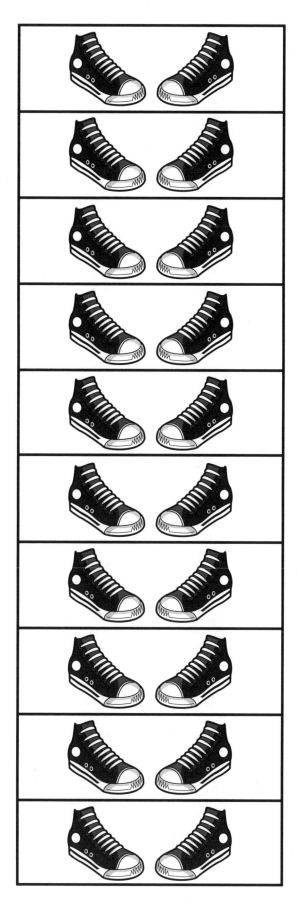

Optional 1-20 Spinner

To use the spinner, place a paper clip in the center. Place pencil point through the paper clip at the center of the spinner. Holding the pencil securely with one hand, spin the paper clip with the other hand.

Basic Place Value

 What is Place Value?

The value of a digit is determined by its position in a number. In our base-ten system, we use ten digits (0-9) to represent all numbers. Place value is the relative worth of a digit based on where it is placed in a number. In the number 482, the first digit is four. It has a value of 400 because it is located in the hundreds place. The second digit is eight. It has a value of 80 because it is located in the tens place. The last digit is two. It has a value of 2 because it is located in the ones place. Place value can be confusing for some students because the same digit represents different values depending on where it is located in a number.

 Formative Assessment

To find out if a student understands place value ask the student to write numbers. Begin with single digit numbers (2, 5, 7, 9). If the student is successful try 2-digit numbers (10, 15, 47, 90, 8 tens and 6 ones, 3 ones and 5 tens). If the student is successful, try 3-digit numbers (184, 269, 420, 503, 3 hundreds and 4 tens and 7 ones, 9 tens and 2 hundreds and 4 ones). If the student is not successful, ask the student to show the numbers with base-ten blocks. Find out what the student knows about place value by uncovering specific gaps in understanding.

Successful Strategies

Place value is a concept that takes time and experience to develop. Encourage students to make numbers with base-ten blocks and identify the amount of tens and ones (and hundreds, if appropriate) that are contained within the numbers. Writing numbers often exposes gaps in understanding. A student may write 400302 to represent the number 432 or 503 to represent the number 53. In these cases, encourage students to show numbers with objects (interlocking blocks are very powerful) and write the digits above or below the model to help them understand place value. Use numbers that are meaningful to students.

> ### Math Words to Use
>
> **place, value, digits, ones, tens, hundreds**

Questions at Different *Levels of Cognitive Demand*

EASY

Recall: *How many tens are in the number?*

Comprehension: *What is the place value of 6 in 64?*

Application: *How could you show the number a different way?*

Analysis: *How would you compare the same digit in different place values?*

Evaluation: *Why is place value an important concept?*

Synthesis: *What if there were nine more tens in the number?*

COMPLEX

Tiny Tens

*A game designed to build the concept
of basic place value*

Materials:

- Tiny Tens Game Board
- Tiny Tens Digit Cards

PLAYERS:
2

Directions:

1. Players mix digit cards and place cards face-down in a pile.
2. The first player draws a digit card from the pile and places the digit card in any box on the left side of the game board.
3. The other player draws a card from the pile and places it in any box on the right side of the game board.
4. The goal is to use digit cards to make the smaller 2-digit number.
5. When a (horizontal) row is complete, players compare the two numbers. The player that created the smaller 2-digit number scores one point.
6. If the two numbers are equal, neither player scores a point.
7. The game ends when all of the digit cards are placed. The winner is the player who scores the most points (total possible points are five).

Content Differentiation:

Moving Back: Place cards in the order in which they are chosen from the pile. Show the 2-digit numbers with base ten blocks.

Moving Ahead: Add boxes to the game board to show 3-digit and/or 4-digit numbers. If the student is ready for six-digit numbers try *Which Place is Best?* in Taylor-Cox, J. (2005), *Family Math Night: Math Standards in Action* (p. 42).

Tiny Tens Game Board

Tiny Tens Digit Cards

1	1	1	2
2	2	3	3
4	4	5	5
<u>6</u>	<u>6</u>	7	7
8	8	<u>9</u>	<u>9</u>

What are Basic Fractions?

Basic fractions are simple common fractions. Fractions are used to name part of a whole or part of a set. The numerator tells how many and the denominator tells the total number of parts. For example, $\frac{1}{4}$ is one out of four.

Formative Assessment

To find out if a student understands basic fractions, ask the student to color specific fractions. Provide the following shapes and say the fraction.

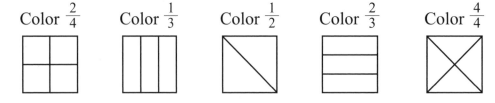

Color $\frac{2}{4}$ Color $\frac{1}{3}$ Color $\frac{1}{2}$ Color $\frac{2}{3}$ Color $\frac{4}{4}$

If the student is successful, try other fractions (through tenths). Present fractions in random order. Identify which fractions students know and which fractions students do not yet know.

Successful Strategies

Use various manipulatives when working with fractions, such as fraction circles, fraction bars, fraction strips. It is important that students do not develop the misconception that all regional fractions are made from only one type of shape. Additionally, present unequal parts to students and encourage them to identify that they are not equal. For example, two unequal parts are not halves.

Math Words to Use

fraction, whole, parts, numerator, denominator, halves, thirds, fourths, fifths, sixths, sevenths, eighths, ninths, tenths

Questions at Different *Levels of Cognitive Demand*

EASY

Recall: *What is the fraction?*

Comprehension: *How many equal parts are there?*

Application: *How could you show how many more parts you need to make one whole?*

Analysis: *Which fraction is greater?*

Evaluation: *How could you order the fractions?*

COMPLEX **Synthesis:** *What is a different way to write fractions?*

My Pie
A game designed to build the concept of basic fractions

Materials:

- **20 chips of one color for each player**
- **My Pie Game Board**
- **2 Dice (1-6 labeled with numerals)**

> **PLAYERS: 2**

Directions:

1. **Roll the dice and use the numbers to make a fraction (1 or less).**
2. **Find the fraction on the My Pie game board. Claim the fraction by covering it with one of your chips.**
3. **Players take turns rolling the dice, making fractions, and claiming fraction pies with chips.**
4. **If a player rolls a fraction that is equal to 1, the player has a choice of which pie to claim.**
5. **If a player rolls a fraction that is claimed by the other player, she may take the other player's chip and claim the pie with her own chip. Taken chips are eliminated from the game. If the player already claimed the pie, her turn is missed.**
6. **The game is over when a player runs out of chips.**
7. **The winner is the player with the most claimed pies on the game board.**

Content Differentiation:

Moving Back: Use fraction circles to make each fraction that is rolled.

Moving Ahead: Look for other equivalent fractions ($\frac{1}{2} = \frac{2}{4} = \frac{3}{6}$). Try writing fractions on index cards as you play. Order the fraction index cards forming a number line.

My Pie Game Board

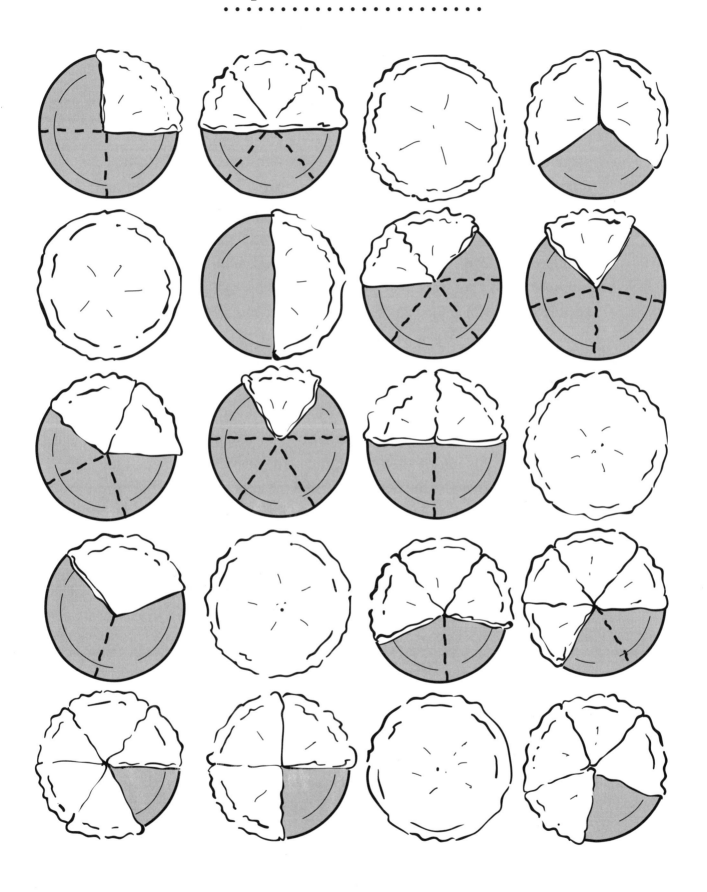

What is Estimation?

Estimation involves approximating quantity. There are four types of estimation; 1) true approximations; 2) overestimation; 3) underestimation; and 4) range-based estimations (Taylor-Cox, 2001). True approximations are typically used with complex numbers. Overestimating and underestimating are commonly situation specific. Range-based estimation involves using an upper end and a lower end to enclose the estimate. We do not want students to think that estimating the exact answer is better than estimating within a reasonable range.

Formative Assessment

To find out if a student can estimate, show the student a pile of 10 red cubes and 30 blue cubes. Tell the student that there are 10 red cubes and ask her to use this information to estimate how many blue cubes. If the student begins to count the cubes, cover up the cubes with a piece of paper and explain that when we estimate we do not count all of the objects. We use the information that we know to make a reasonable guess. If the student is successful, try larger quantities. If the student is not successful, try smaller quantities. Notice if the student estimates using an exact number or if the student offers a range as the estimate.

Successful Strategies

Estimating quantity is best understood through the use of a benchmark (also called referent or anchor). Students use "what they know" to estimate what is unknown. Benchmarks can be more or less than the unknown amount. To help students increase skill in estimating quantity, we need to provide many experiences with estimating and teach students to use what they know about the quantity.

> **Math Words to Use**
>
> **estimation, estimate, approximation, benchmark, referent, anchor**

Questions at Different *Levels of Cognitive Demand*

EASY

Recall: *What is estimation?*

Comprehension: *How do you estimate?*

Application: *How did you use ten in your estimation?*

Analysis: *How would you compare the different estimates?*

Evaluation: *Is your estimate within a reasonable range? Why or why not?*

Synthesis: *What if all estimations need to be over to have enough?*

COMPLEX

Clover Patch

A game designed to build the concept of estimation

Materials:

- **Clover Patch Spinner**
- **Paper clip and pencil**
- **Clover Patch Score Sheet**
- **Copies of the Clover Patch Game Board for each player**

PLAYERS: 2

Directions:

1. **The first player spins the Clover Patch spinner. If the player spins one finger, he gives a range of one (example 20-21) to estimate how many clovers his finger will touch when he places it on the Clover Patch game board.**

2. **The player places his finger anywhere on the clover patch and uses a pencil to trace his finger. The player counts to find out how many clovers his finger actually touched. If the actual number of clovers falls within the range, the player scores one point.** *NOTE: If the player spins two fingers, he gives an estimate in a range of two (example 20-22). If the player spins three fingers, he gives an estimate in a range of three (example 20-23). If the actual number of clovers falls within the range given, the player scores the corresponding number of points (1 point for one finger, 2 points for two fingers, 3 points for three fingers). Clovers can only be touched one time.*

3. **Players keep a running total on the score sheet.**

4. **The winner is the first player to score 10 points or more.**

Content Differentiation:

Moving Back: **Play the game without the spinner. Players take turns using one finger.**

Moving Ahead: **Spin twice and add the fingers together to increase the quantity. Tape together two copies of the Clover Patch game board.**

Clover Patch Spinner & Score Sheet

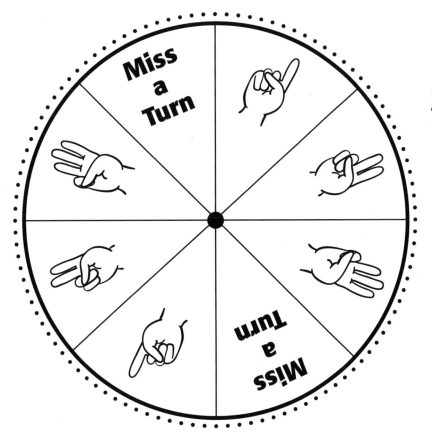

To use the spinners, place a paper clip in the center. Place pencil point through the paper clip at the center of the spinner. Holding the pencil securely with one hand, spin the paper clip with the other hand.

Name:_____	Name:_____

Clover Patch Game Board

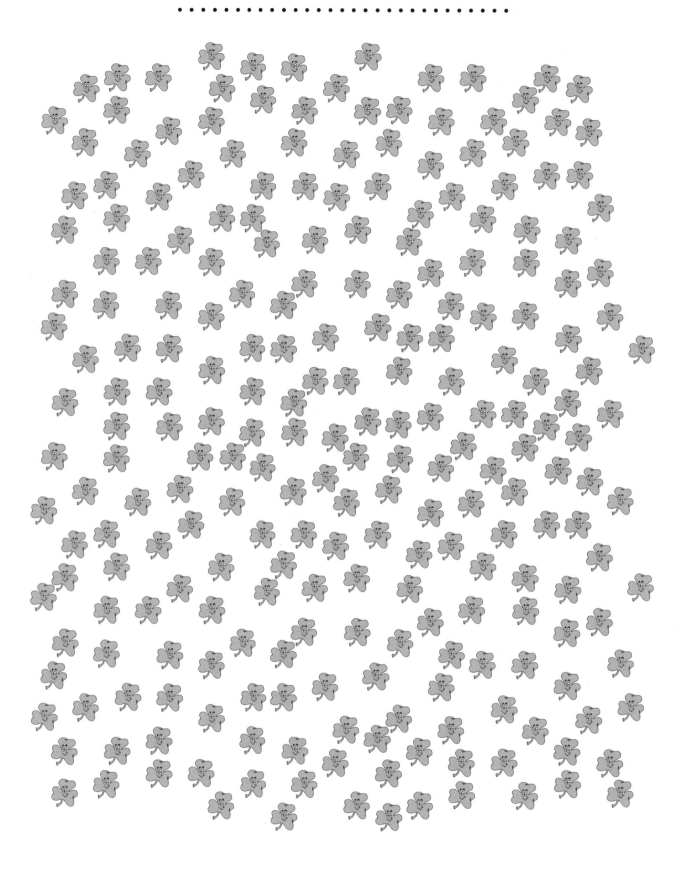

Math Intervention: Grades PreK-2 • *Jennifer Taylor-Cox* • *© Eye On Education*

CHAPTER

Addition and Subtraction Concepts

Total and Parts
Counting On and Counting Back
Joining Sets
Number Line Proficiency
Take Away Subtraction
Missing Part Subtraction
Comparison Subtraction
Adding and Subtracting Tens
Adding Doubles and Near Doubles
Fact Families—Addition and Subtraction
Partial Sums
Partial Differences
Near Tens for Addition and Subtraction
Equal Differences

Total and Parts

What is the Total and Parts Concept?

The total and parts concept encourages students to see the parts of whole numbers. By focusing on total and parts, students learn about addition and subtraction. They begin to understand that the parts make the total and that the total is made of parts. The beauty of teaching total and parts is that we can present early addition and subtract concepts at the same time rather than teaching these operations separately.

Formative Assessment

To find out if a student understands total and parts, ask the student to describe the total and parts of the following trains of cubes;

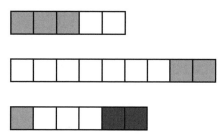

Notice if the student provides the total and parts in the descriptions. If the student is successful, try different parts and larger totals. Identify which totals and parts the students know and which they do not know.

Successful Strategies

Interchange addition and subtraction ideas as students work with total and parts. Encourage students to read aloud and record the parts and total. Using different colors to show the parts often helps students better identify the parts. Grouping the parts is also beneficial. As students gain comfort with total and parts begin working with more than two parts of each total.

> ### Math Words to Use
>
> **total, whole, parts,
> addition, subtraction**

 ## Questions at Different *Levels of Cognitive Demand*

EASY

Recall: *What is the total?*

Comprehension: *How do the parts relate to the total?*

Application: *How could you show different parts of the total?*

Analysis: *How could you compare the parts of more than one total?*

Evaluation: *What helps you understand the parts of a total?*

COMPLEX **Synthesis:** *What if the total had many parts?*

Big Top Twelve

A game designed to build the concept of total and parts

Materials:

- Die (labeled 1-6 with numerals)
- Crayons or colored pencils
- Big Top Twelve Game Board for each player

PLAYERS: 2 OR MORE

Directions:

1. The first player rolls the die. The number showing is one part of twelve.

2. The player decides which Big Top she wants to use to show the part. The player colors the part with a specific color. For example, if the player rolls five, she colors five rectangles (next to one another) in one of the Big Tops with a black crayon.

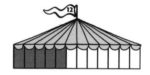

3. The next player rolls and colors the part of twelve on her game board.

4. Players take turns rolling and coloring parts of twelve. Each new part of the Big Top is shown using a different color.

5. The number rolled cannot be split. If the number rolled cannot be played, the player loses her turn.

6. The first person to complete all three Big Tops is the winner.

Content Differentiation:

Moving Back: Try Big Tops that are less than 12.

Moving Ahead: Play Big Top Twelve without the game board. Each time a number is rolled, it becomes part of one of three addition equations. For example, $2 + 3 + 5 + 2 = 12$. Try other Big Top values.

Big Top Twelve Game Board

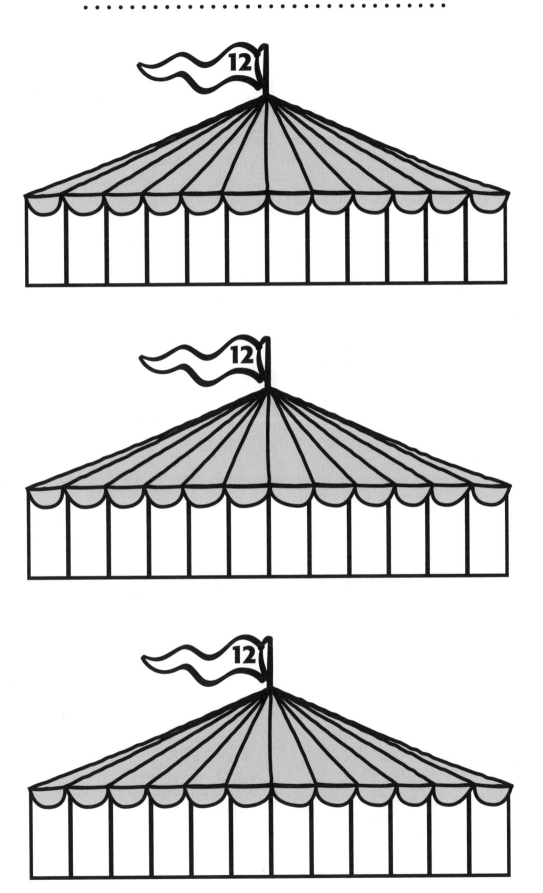

Big Top Twelve Game Board

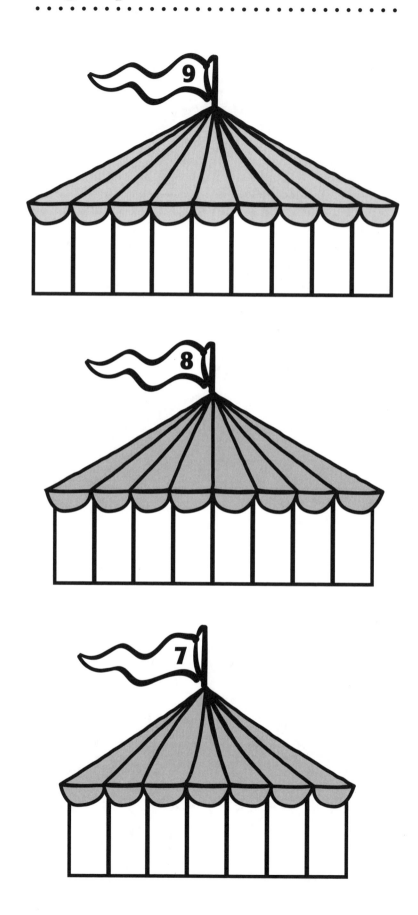

Big Top Twelve Game Board

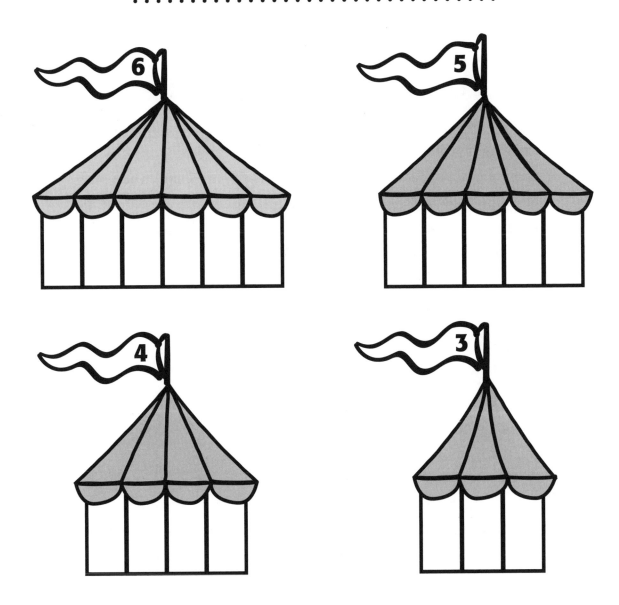

CONCEPTS:
Counting On and Counting Back

 ## What are Counting On and Counting Back?

Counting on and counting back are strategies that help students with addition and subtraction. Counting on and counting back are efficient strategies when the addition equation is +1, +2, or +3 and when the subtraction equation is -1, -2, or -3. In a situation such as 48 + 23, counting on by ones is not the most efficient strategy because it is too time consuming. Students need to know how to count on and count back, but they also need to judge when it is a good idea to count on or count back.

 ## Formative Assessment

To find out if a student can count on and count back ask the student to name one more than several given numbers (23, 57, 79). Then try two more, three more, one less, two less, and three less.

23 and one more	57 and one more	79 and one more
23 and two more	57 and two more	79 and two more
23 and three more	57 and three more	79 and three more
One less than 23	One less than 57	One less than 79
Two less than 23	Two less than 57	Two less than 79
Three less than 23	Three less than 57	Three less than 79

Identify which situations the students answer correctly and how quickly the students respond to determine how comfortable the students are with counting on and counting back by ones.

Successful Strategies

Identifying when it is efficient and when it is not efficient to use counting on and counting back is just as important as knowing how to count on and count back. When students know how and when to count on by ones, we can introduce counting on by numbers other than one. As discussed previously, counting on by ones is not an efficient strategy to use to solve $48 + 23$; although it would be efficient to count on by tens and then ones to solve the equation ($48 + 23 = 48 + 10 + 10 + 1 + 1 + 1$). Similarly, students can count on by other numbers when they are ready.

Math Words to Use

count on, count back, one more, two more, three more, one less, two less, three less

---

 ## Questions at Different *Levels of Cognitive Demand*

EASY

Recall: *What is one more?*

Comprehension: *What does counting back mean?*

Application: *When is it a good idea to count on by ones?*

Analysis: *How are counting on and counting back similar?*

Evaluation: *When and why is counting back a good strategy?*

Synthesis: *How could you use skip counting by fives to count on?*

COMPLEX

Nifty Fifty

*A game designed to build the concepts
of counting on and counting back*

Materials:

- **Die (labeled 1, 1, 2, 2, 3, 3)**
- **Nifty Fifty Score Sheet for each player**

PLAYERS:
2 OR MORE

Directions:

1. **All players record 50 in the "start" column of the first round on their score sheets.**
2. **The first player rolls the die and records the rolled number on his score sheet. The player chooses whether to count on or count back the number rolled and records the "end" number.**
3. **The next player rolls, records on her score sheet, and counts on or counts back to find the "end" number.**
4. **The "end" number becomes the start number for each new round.**
5. **The player who has 50 after the 10th round is the winner.**
6. **If no players have 50 after the 10th round, the player who is closest to 50 is the winner.**

Content Differentiation:

Moving Back: **Change the start/goal number to 20. Write the equation for each round.**

Moving Ahead: **Play Nifty Fifty without the score sheet. Mentally keep track of "start" and "end" numbers. Try greater start/goal numbers or add more rounds.**

Nifty Fifty Score Sheet

Round	Start	Rolled	End
1st			
2nd			
3rd			
4th			
5th			
6th			
7th			
8th			
9th			
10th			

What is the Joining Sets Concept?

Joining sets is an addition concept. Joining sets involves combining quantities and finding the total. Each set that we join is considered an addend. The total of the combined sets is the sum. For some students, the introduction of symbols (+, =), equation formats, and rules come before they have established an understanding of what it means to join sets. These students often struggle with addition because they are trying to remember symbols, formats, and rules that are meaningless to them. It is critical that we help students develop a true understanding of joining sets so that the symbols, formats, and rules make sense to them.

Formative Assessment

To find out if a student can join sets use the following number stories. Ask students to explain how they figured out each answer with words, pictures, or models.

Michael has three cookies. Carlos gives him two more cookies. How many cookies does Michael have now?

Gloria has large books and small books on her shelf. There are four large books and five small books. How many books are on Gloria's shelf?

Ten girls line up. Four boys line up. How many students are in the line?

Successful Strategies

To help build students' conceptual knowledge, provide them with models and meaningful situations. It is important that they visualize, manipulate, and talk about joining sets. Teachers do not need to be the sole providers of number stories. Students can and should create story problems as they learn what it means to join sets. Additionally, students need to understand that many sets can be joined. Misconceptions may occur for students if we only provide addition problems that have two addends.

> ## Math Words to Use
>
> **join, combine, add,
> addition, plus, sets,
> quantities, amounts**

  ## Questions at Different *Levels of Cognitive Demand*

EASY

Recall: *What is four and three more?*

Comprehension: *How would you summarize what it means to join sets?*

Application: *How would you show four plus three?*

Analysis: *How are 2 + 3 and 4 + 1 alike?*

Evaluation: *What is the best way to join these sets?*

COMPLEX   **Synthesis:** *How could you rearrange 5 + 2 + 5 + 3?*

Domino Dozen

A game designed to build the concept of joining sets

Materials:

- Set of Double-Six Dominoes

PLAYERS: 2

Directions:

1. Place dominoes face down so that dots cannot be seen. To start the game, flip one domino at random.
2. The first player flips a domino and joins the sets of dots on both dominoes.
3. If the total number of dots is a dozen (twelve), the player takes both dominoes. If the total number of dots is not a dozen, the player leaves the dominoes face up.
4. The second player flips one domino and looks for any face-up domino that could be combined with her domino to make a dozen dots. If a combination of twelve can be made, the player takes the dominoes. If a combination of twelve cannot be made the player leaves the dominoes face up.
5. Play continues until all dominoes have been flipped over.
6. The winner is the player with the most dominoes.

Content Differentiation:

Moving Back: Use cubes to represent the number of dots on the dominoes. Or focus on joining the sets of dots on individual dominoes, looking for combinations of six.

Moving Ahead: Write equations for each flip, for example. Or look for combinations of 18 using double-nine dominoes.

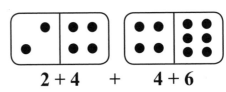

2 + 4   +   4 + 6

Number Line Proficiency

What is Number Line Proficiency?

Number line proficiency involves understanding the order and spacing of numbers assigned to points on a line. Even though every point on a number line has a unique real number (including fractions of fractions), students need to begin with number lines that have the counting by ones points labeled before moving to more complex labels. Number line proficiency includes knowing that on a horizontal number line, movement to the right represents addition and movement to the left represents subtraction. In addition to horizontal number lines, students should also have experiences with vertical number lines. Experiences with number lines that are separated into line segments, such as hundred charts and calendars, and circular number lines (technically not lines), such as analog clocks are also beneficial. Using number lines helps students build understanding of computation and number relationships.

Formative Assessment

To find out if a student has number line proficiency, show parts of various number lines and see if students can fill in the missing numbers. Notice if the student quickly identifies the correct missing numbers. Look for patterns in strategies and in errors.

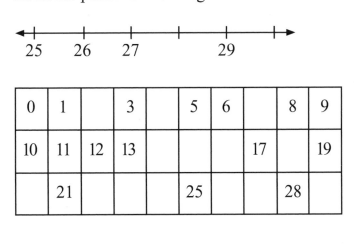

Successful Strategies

For some students, the process of cutting out the rows of a hundred chart and taping the ends together helps them build understanding. They see how the number line is separated into line segments on the hundred chart. This same process can also be used with the calendar.

Math Words to Use

number line, order, count, before, after

 ## Questions at Different *Levels of Cognitive Demand*

EASY

Recall: *What number comes next?*

Comprehension: *What are some examples of numbers that come before?*

Application: *How could you use the number line to show this math situation?*

Analysis: *What numbers go between other numbers on the number line?*

Evaluation: *Which number line works best? Why?*

Synthesis: *How would you design a number line that overlaps with another number line?*

COMPLEX

Number Line Fill Up

A game designed to build the concept of number line proficiency

PLAYERS: 2 OR MORE

Materials:

- **Number Line Fill Up template for each player**
- **Decahedron die (ten-sided, labeled 0-9) or 0-9 spinner** *(see page 62)*

Directions:

1. **Each player rolls the decahedron die (or spins the 0-9 spinner) one time to find his start number.**
2. **Players decide where to write their start numbers on the Number Line Fill Up template.**
3. **All other rolls (or spins) are numbers shared by all players.**
4. **Each player records the new numbers one at a time. In this game, players are rolling (or spinning) one digit at a time to make 2-digit numbers in a counting by ones sequence. If a number cannot be used, the player waits for the next roll (or spin). Once a number is written, it cannot be moved.**
5. **The winner is the first player to complete an accurate number line chunk including four 2-digit numbers in order.**

Content Differentiation:

Moving Back: **Show values with base ten blocks or cubes as students build the number line, or try the game using single-digit numbers.**

Moving Ahead: **Add a third section to each number box and try the game with 3-digit numbers, or try a skip counting sequence on the number line.**

Number Line Fill Up

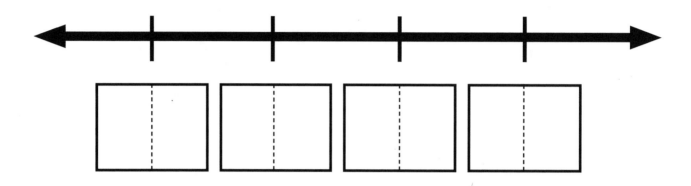

Number Line Fill Up

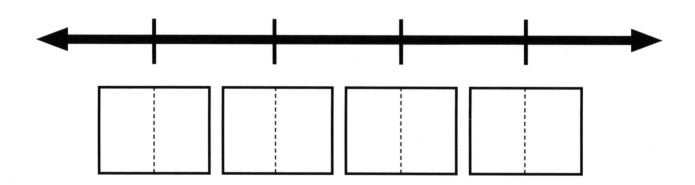

What is the Take Away Subtraction Concept?

Take away subtraction is one of three types of subtraction. Take away subtraction involves set removal. In number stories, a certain amount is removed, eaten, given, lost, hidden, or otherwise taken away. In a subtraction equation, the subtrahend is being subtracted from the minuend to find the difference. In the example $7 - 2 = 5$, 7 is the minuend, 2 is the subtrahend, and 5 is the difference. Students need many meaningful experiences to build conceptual understanding of take away subtraction.

Formative Assessment

To find out if a student understands take away subtraction, ask the student to tell a story that matches the following equation; $10 - 4 = 6$. Notice if the student uses words that represent take away subtraction. If the student is not successful, try a smaller equation such as $4 - 1 = 3$ and invite the student to use manipulatives to demonstrate as he tells the number story. If the student is successful with $10 - 4 = 6$, give a larger equation such as $25 - 6 = 19$ and invite the student to write the equation and tell a related number story.

Successful Strategies

The take away subtraction concept is best understood in context. Students need to engage in many types of take away situations to build conceptual understanding. Our goals are not for students to memorize vocabulary such as minuend and subtrahend. Instead we need to model these words as we teach the students the true meaning of take away subtraction.

> **Math Words to Use**
>
> subtract, subtraction, take away, minus sign, equation, minuend, subtrahend, difference

 ## Questions at Different *Levels of Cognitive Demand*

EASY

Recall: *What is eight take away three?*

Comprehension: *How would you explain take away subtraction?*

Application: *How would you use take away subtraction in this situation?*

Analysis: *Which number story represents take away subtraction?*

Evaluation: *Which type of subtraction best represents this math situation?*

Synthesis: *If you started with 6 more, how would that affect the difference?*

COMPLEX

Hungry Frog

A game designed to build the concept of take away subtraction

Materials:

PLAYERS: 2 OR MORE

- **Hungry Frog Game Board**
- **40 small counters**
- **Dice (two six-sided cubes labeled 1-6 with numerals)**

Directions:

1. Players place one counter on each fly on the game board.
2. The first player rolls both dice and decides which number she wants to use to represent how many flies Hungry Frog will eat. The player tells a number story about Hungry Frog eating flies, models the take away subtraction by taking away counters, and states the subtraction equation. For example, there are 40 flies. Hungry Frog eats three. Now there are 37 flies left, $40 - 3 = 37$.
3. The next player rolls the dice and chooses one of the numbers to be the subtrahend. The player tells the new subtraction story and states the equation. For example, $37 - 5 = 32$.
4. Players take turns rolling the dice, choosing a subtrahend, telling and modeling the associated number story, and stating the equation.
5. The winner is the player who has Hungry Frog eat the last fly. An exact number must be rolled to win.

Content Differentiation:

Moving Back: Use half of the game board and start with 20 flies.

Moving Ahead: Use mental math to play this game without counters or the game board. Tell different kinds of number stories.

Hungry Frog Game Board

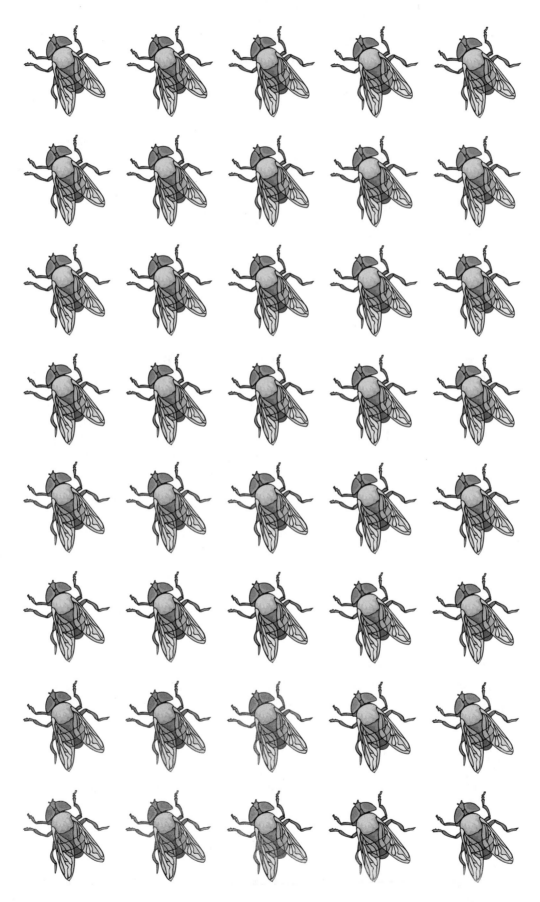

Missing Part Subtraction

What is Missing Part Subtraction?

Missing part subtraction is one of three types of subtraction. Missing part subtraction involves an unknown part of the whole. Either the subtrahend or the difference is unknown. For example, $10 - \square = 2$ or $10 - 8 = \square$. Missing part subtraction is often used to solve start unknown or change unknown addition problems. For example, $\square + 2 = 10$ or $8 + \square = 10$. When working with missing part subtraction, it is important that students use "subtract" rather than "take away" to more accurately describe the mathematics situation.

Formative Assessment

To find out if a student understands missing part subtraction, ask the student to solve the following number stories;

Melanie has 5 bracelets in her jewelry box. She puts some of the bracelets on her wrist. There are 2 bracelets left in her jewelry box. How many bracelets are on her wrist?

An apple tree has 11 beautiful apples. The wind blows 4 apples off the tree. How many apples are still on the tree?

Lenny bakes 24 cupcakes. He packs a lot of the cupcakes in a box and leaves 6 on the table. How many cupcakes are in the box?

Successful Strategies

To help students understand the concept of missing part subtraction use models to show the parts of the total. Color coding parts and arranging the parts in groups are helpful ways to highlight the parts within the whole set.

> ### Math Words to Use
>
> **subtract, subtraction, difference, minuend, subtrahend, minus sign, missing part**

  ## Questions at Different *Levels of Cognitive Demand*

EASY

Recall: *What is 6 – 2?*

Comprehension: *How do you know that 10 – 3 = 7?*

Application: *Is there another number story that could go with this subtraction equation?*

Analysis: *Which number stories represent missing part subtraction?*

Evaluation: *What is the best way to model missing part subtraction?*

Synthesis: *How could you reorganize these facts to show a new pattern?*

COMPLEX

Broken Towers

A game designed to build the concept of missing part subtraction

Materials:

- Linking cubes (20 for each player)

PLAYERS: 3 OR MORE

Directions:

1. The first player states a number (1-20).
2. All players build a tower using the stated number of cubes.
3. When the towers are completed, each player holds her tower under the table. The players break their towers into two parts.
4. One at a time, players show one part of their tower by placing it on the table. The other players count the cubes in the part of the tower that is showing and try to name how many cubes are in the missing part under the table.
5. The first player to name the correct number of cubes in the missing part of the tower earns one point. The player showing the tower can also earn one point by successfully stating the missing part subtraction equation.
6. After all players have shown their towers, the round is complete. A new number is given for the next round.
7. The winner is the player with the most points after five rounds.

Content Differentiation:

Moving Back: **Build towers of ten or less.**

Moving Ahead: **Break towers into three parts.**

Comparison Subtraction

What is Comparison Subtraction?

Comparison subtraction is one of three types of subtraction. In comparison subtraction, the focus is on the difference between quantities. Comparing means finding out how many more or how many less one number is compared to another number. As with missing part subtraction, comparison subtraction is not "take away." Finding the difference does not involve removal; therefore we should call it subtraction rather than "take away." Differences are reported in absolute values. This means that we do not need to teach students rules such as *the big number goes first when we subtract*. Instead, we can focus on teaching the concept of comparison subtraction because the difference between two numbers can be positive or negative.

Formative Assessment

To find out if a student understands comparison subtraction, give the student some numbers to compare. Ask the student to explain how he solved each problem.

What is the difference between 5 and 4?
What is the difference between 10 and 7?
What is the difference between 15 and 9?
What is the difference between 20 and 12?

If the student is successful, try numbers with greater differences. If the student is not successful, encourage the student to use objects to model the differences.

Successful Strategies

Comparison subtraction is difficult for many students. To help highlight differences between quantities, students can use manipulatives. Towers of cubes are especially helpful. When we ask the student to name the difference, we can discuss how many more are needed to make the towers even or equal.

> ## Math Words to Use
>
> **subtract, subtraction, compare, comparison, difference, subtrahend, minuend, minus sign**

Questions at Different *Levels of Cognitive Demand*

EASY

Recall: *What is the difference between 15 and 5?*

Comprehension: *How would you explain comparison subtraction?*

Application: *How would you show 12 − 4 with pictures?*

Analysis: *Which number stories represent comparison subtraction?*

Evaluation: *What is the best way to find the difference? Why is it best?*

Synthesis: *How would you design a new template for comparison subtraction?*

COMPLEX

Compare Bear

A game designed to build the concept of comparison subtraction

PLAYERS: 2

Materials:

- **Number Mat for each player**
- **Two different colored bear counters for each player**

Directions:

1. **The first player drops the two bears onto the Number Mat. The player announces the two numbers that the bears land on (or are closest to) and states the difference. For example, if the bears land on 14 and 8, the player says, "The difference between 14 and 8 is 6."**
2. **The player also tells a comparison subtraction story. For example, "Blue Bear has 14 cookies. Red Bear has 8 cookies. Blue Bear has 6 more cookies than Red Bear."**
3. **The next player drops two bears on his number mat. The player announces the two numbers, states the difference, and tells a comparison subtraction story.**
4. **The player with the greater difference scores one point.**
5. **Players take turns dropping bears on their number mats, stating the difference, and telling the associated number stories.**
6. **The winner is the first player to score ten points.**

Content Differentiation:

Moving Back: **Use connecting cubes to show the numbers to help students see the differences.**

Moving Ahead: **Try the same game using a hundred chart as the number mat.**

Number Mat

· · · · · · · · · · · · · · · · · · · ·

1	2	3	4	5
6	7	8	9	10
11	12	13	14	15
16	17	18	19	20

Adding and Subtracting Tens

What is the Adding and Subtracting Tens Concept?

Adding and subtracting tens builds fluency with number relationships and computation. When students understand the adding and subtracting tens concept, they are more apt to use mental math to solve problems and verify answers. Initially, students add and subtract tens to and from multiples of ten. Building on this idea, they learn to add and subtract tens to and from any number.

Formative Assessment

To find out if a student can add and subtract tens, ask the student to solve the following equations;

20 + 10 =

40 − 10 =

50 + 30 =

70 − 20 =

65 + 10 =

82 − 10 =

38 + 40 =

97 − 30 =

Ask the student to explain how he solved each problem. Analyze the student's proficiency through his accuracy, speed, and explanations.

Successful Strategies

One of the foundational concepts associated with adding and subtracting tens is rational counting by tens. Encourage students to skip count by tens beginning with zero and with numbers other than zero. To further help students understand the adding and subtracting tens concept, use base ten blocks or other manipulatives to model the computation.

> ### Math Words to Use
>
> **tens, add, addition, subtract, subtraction, minus sign, plus sign**

Questions at Different *Levels of Cognitive Demand*

EASY

Recall: *What is 53 + 10?*

Comprehension: *How do you add ten to a number?*

Application: *How would you use 92 – 10 to solve 92 – 20?*

Analysis: *Which facts go together? Why?*

Evaluation: *How does skip counting help us add and subtract ten?*

Synthesis: *How could you use the adding and subtracting tens concept to add and subtract nines?*

COMPLEX

Hundred Chart Tic-Tac-Toe

A game designed to build the concept of adding and subtracting tens

Materials:

PLAYERS: 2

- Markers or crayons
  (different color for each player)
- Paper

Directions:

1. Each player decides which color marker/crayon to use.
2. One player draws a tic-tac-toe board on the paper.
3. The first player writes a 2-digit number in one of the spaces.
4. The next player uses adding or subtracting ones or tens to write a related number in one of the spaces. The placement of numbers is relative to position on a hundred chart.
5. Players take turns filling in related numbers.
6. The winner is the first player to have three numbers in a horizontal, vertical, or diagonal row. For example;

54	55	56
64	65	66
74		76

Content Differentiation:

Moving Back: Use a hundred chart as a reference.

Moving Ahead: Make a tic-tac-toe board with more lines and try four in a row.

Adding Doubles and Near Doubles

What is the Adding Doubles and Near Doubles Concept?

There is something uniquely fascinating about doubles. Often students are keenly interested in doubles, which propels them to master these facts earlier than other facts. The smaller doubles (1+1, 2+2, 3+3, 4+4, 5+5) are often learned quickly. Larger doubles (6+6, 7+7, 8+8, 9+9) take more time for some students. To help students learn doubles, provide many experiences with doubling situations. Students also need to conceptually understand addition and the idea of "twice as many." When students own their doubles, meaning they do not need to count all or count on, they are ready to learn how to use this knowledge to solve near double addition problems. When teaching the near doubles concept, encourage students to identify the two doubles facts that the near double falls between and allow them to use plus one or minus one to solve the equation. For example, $7 + 8 = 15$ falls between $7 + 7 = 14$ and $8 + 8 = 16$. Rather than providing rules for students to follow, allow the students to decide which double they want to use to solve the near double problem. In this way, students learn more about number relationships. They know that they need to add one or subtract one because they understand the math involved.

Formative Assessment

To find out if a student can add doubles and near doubles, provide addition situations and ask them how they solved the problems. Ask the student to solve $5 + 5$ and see how quickly the student responds. Try other doubles ($1 + 1$ through $10 + 10$). Analyze the level of ownership by the accuracy and speed in which the student solves the problems. If the student owns her doubles, provide near doubles problems and ask students to explain how they solved the problems. Identify which doubles and near doubles the student owns and which the student does not own.

Successful Strategies

Teaching students to identify doubles and near doubles is the first step in ownership of these facts. Ask students to explain which two doubles facts "sandwich" the near doubles fact. Use manipulatives (towers of cubes or stacks of counters) to model one more and one less. Write out the facts in order so that students can see the patterns. As students gain ownership of doubles and near doubles, we need to encourage them to use this knowledge with related subtraction facts.

Math Words to Use

doubles, twice as many, add, addition, facts, sum, near doubles, one more, one less

 ## Questions at Different *Levels of Cognitive Demand*

EASY

Recall: *What is 4 + 4?*

Comprehension: *How would you explain double?*

Application: *How could you use 6 + 6 to solve 6 + 7?*

Analysis: *Why is 5 + 5 similar to 10 + 10?*

Evaluation: *How would you help someone else solve 7 + 8?*

Synthesis: *What if you were to invent a new way to organize the doubles facts?*

COMPLEX

X Marks the Spot

***A game designed to build the concept
of adding doubles and near doubles***

Materials:

**PLAYERS:
2**

- 0-9 Number Cards or 0-9 Number Tiles in a paper cup
- X Marks the Spot Game Boards
- Two highlighters or light markers (different colors)

Directions:

1. Players decide which color highlighter each person will use.
2. Without looking into the cup, the first player chooses a number. The player doubles the number and looks for the sum on the game board.
3. The player marks the sum with a dot using the highlighter to "hold" the spot and places the number back into the cup.
4. The next player chooses a number, doubles the number and looks for the sum on the game board. The player can mark the sum with a dot to hold it OR the player can "claim" any sum that is currently being held with a dot.
5. If the player claims a held sum, the highlighter is used to draw X on the sum.
6. Players take turns choosing numbers, doubling, and holding or claiming sums. *NOTE: All sums must be held before they are claimed. Held sums can be claimed by any player.*
7. The winner is the first player to have four claimed sums (Xs) in a row (horizontal, vertical, or diagonal).

Content Differentiation:

Moving Back: Play the game without holding sums prior to claiming sums.

Moving Ahead: Play the same game using Near Doubles. The near double is one more or one less than the closest double.

0-9 Number Cards

• • • • • • • • • • • • • • • • • • • •

0	1	2	3	4
5	6	7	8	9

X Marks the Spot Game Board for Doubles

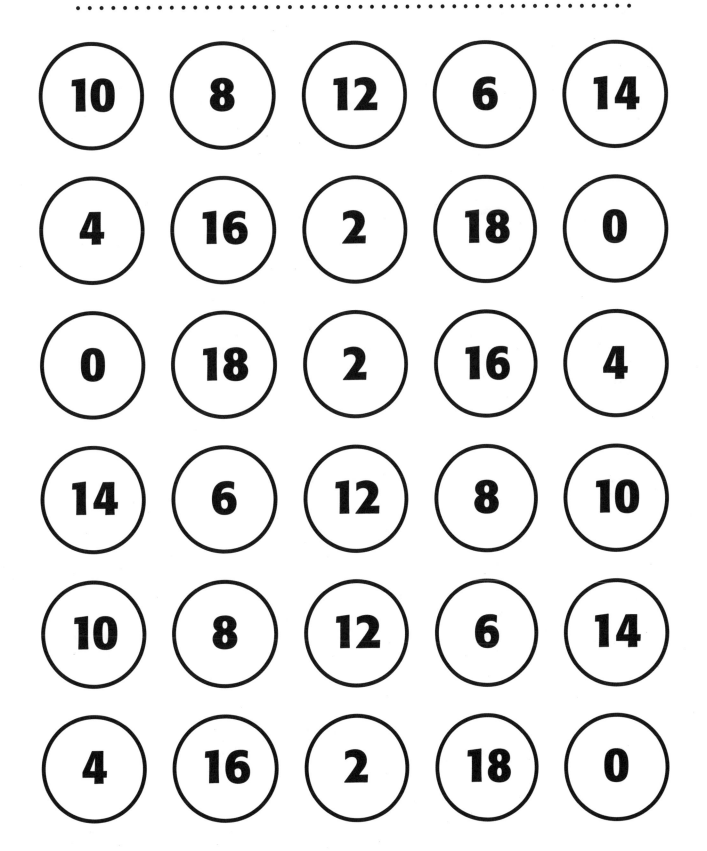

X Marks the Spot Game Board for Near Doubles

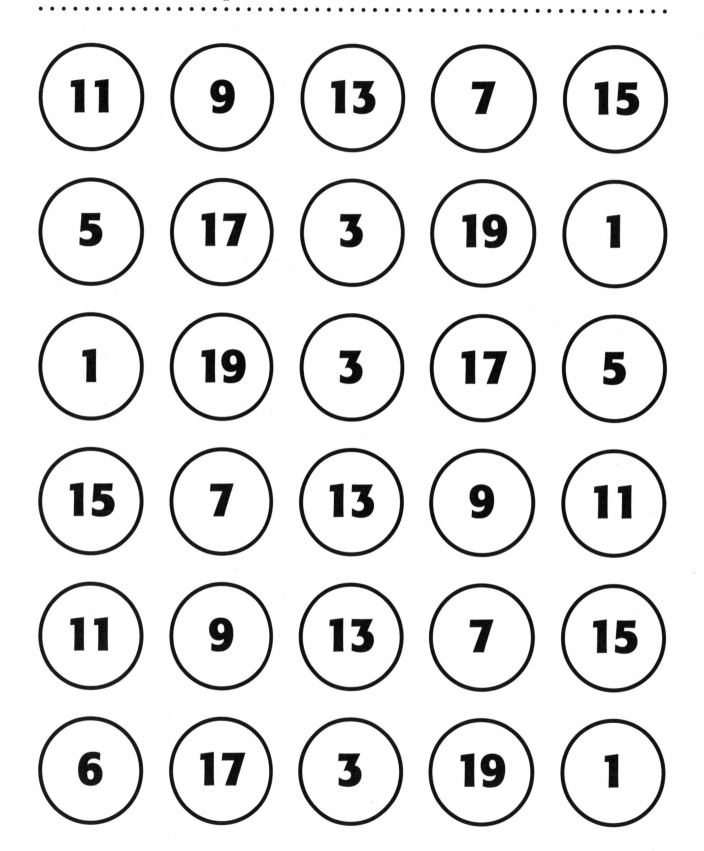

Fact Families—Addition and Subtraction

 What are Addition and Subtraction Fact Families?

Fact families for addition and subtraction are facts that are directly related. In a fact family, there are two addition problems and two subtraction problems. The addition facts are manifestations of the commutative property. The subtraction facts are inverse operations of the addition facts.

 Formative Assessment

To find out if a student understands addition and subtraction fact families, provide the following incomplete fact family and ask the student which fact is missing from the family.

$8 + 4 = 12$

$12 - 4 = 8$

$12 - 8 = 4$

If the student is not successful, try a smaller fact family. If the student is successful, ask her to generate all of the facts in the 7, 8, 15 fact family.

Successful Strategies

Some students struggle with fact families. One of the ways to help these students is to go back to the parts and total concept. Focusing on the parts of a given total encourages students to think about the related addition and subtraction facts. Additionally, help student see why doubles produce identical facts.

Math Words to Use

addition, subtraction, equation, facts, fact families

 ## Questions at Different *Levels of Cognitive Demand*

EASY

Recall: *What is 5 + 2? What is 2 + 5?*

Comprehension: *How would you explain fact families?*

Application: *How would you show the fact family of 6, 7, 13?*

Analysis: *What relationship do you see between 8 + 9 and 17 − 9?*

Evaluation: *What advice would you give to someone who does not understand fact families?*

Synthesis: *Is it possible to have a fact family for this*

COMPLEX *equation 2 + 3 + 4 = 9?*

Funny Bunny

A game designed to build the concept of addition and subtraction fact families

Materials:

- **Funny Bunny Resource Pages**
- **Straws**
- **Tape**

PLAYERS: 2

Directions:

1. Cut out bunnies and tape each bunny on a straw.
2. Place all bunny puppets face down on the table.
3. The first player chooses a bunny and folds back one of the bunny's ears.
4. The other player names the hidden number on the bunny's ear. The player receives one point for each fact in the fact family that is named correctly.
5. Players take turns choosing bunnies, folding ears, and naming the facts in the fact family.
6. The first player to reach 20 or more points is the winner.

Content Differentiation:

Moving Back: Look at the bunny with both ears up and name the facts in the fact family.

Moving Ahead: Use the blank bunnies to make different fact families.

Funny Bunny Resource Page

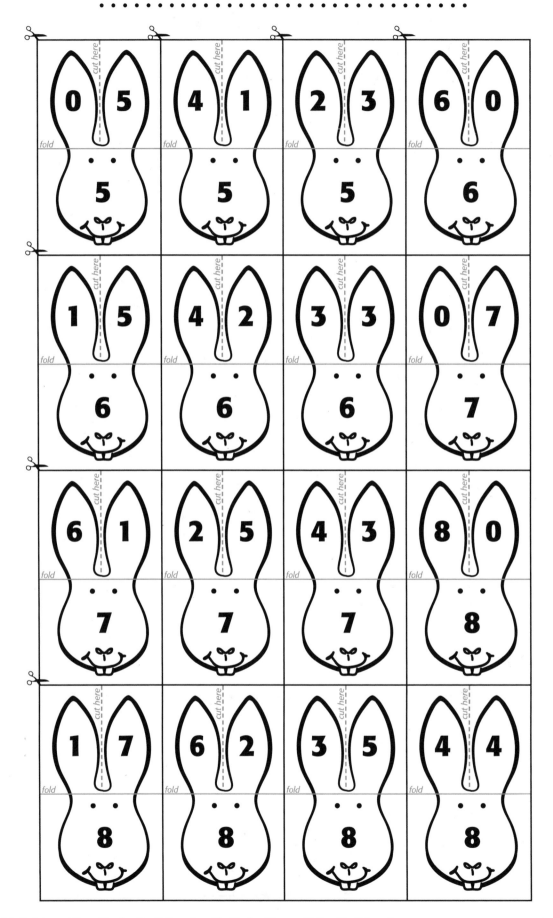

Funny Bunny Resource Page

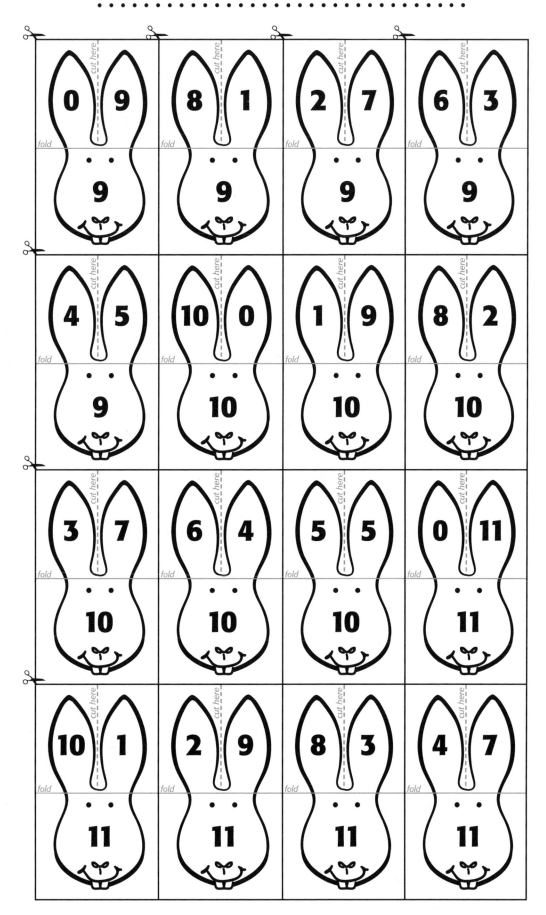

Funny Bunny Resource Page

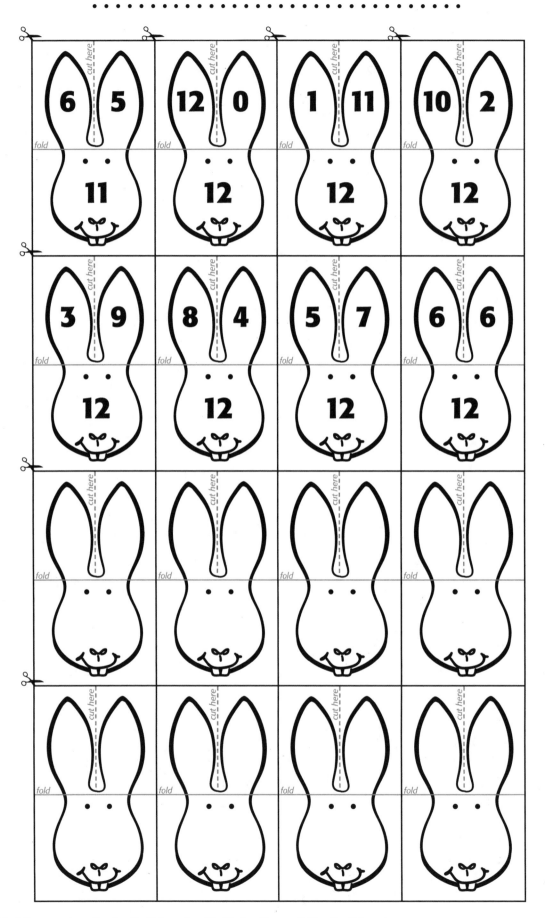

What is the Partial Sums Concept?

The partial sums concept is an addition strategy that involves combining addends into workable clusters. The traditional algorithm for addition involves following procedures, such as *aligning digits in columns* and *always starting with the ones*. The rules of the traditional algorithm are meaningless for many students. They cannot remember what to do because they do not understand the math. Alternative algorithms such as partial sums build conceptual understanding. With partial sums, it does not matter if columns are aligned or if the student starts with the ones or not. There is no need to "carry." For example, $37 + 48 = 70 + 15$. In this situation, the tens are clustered (70 is a partial sum) and the ones are clustered (15 is a partial sum). The partial sums are easily combined into 85. The idea with partial sums is to cluster the addends in a way that invites mental math.

Formative Assessment

To find out if a student understands the partial sums concept, ask the student to solve the following equations;

$23 + 42 =$	$26 + 23 =$	$31 + 39 =$
$68 + 12 =$	$57 + 38 =$	$76 + 17 =$

Ask the student to explain how she solves each problem. If the student uses the traditional algorithm (successfully or not successfully), ask the student if she knows another way to solve the equations. Invite the student to try the problems without aligning the digits in columns and without starting with the ones.

Successful Strategies

Some of the reasons students struggle with the traditional algorithm include the phrases that they have heard. For example, many people say things like, "Carry the one," when they are actually regrouping a ten. Because the math language is not accurate, the students develop misconceptions. The partial sums concept is embedded in place value and conceptual understanding, which help remedy algorithmic misconceptions. Even if students successfully use the traditional algorithm, learning the partial sums concept invites students to strengthen conceptual understanding, increase mental math capacity, and improve flexibility and fluency.

Math Words to Use

partial, part, sum, addition, add, strategy, equation, equal

 ## Questions at Different *Levels of Cognitive Demand*

EASY **Recall:** *What is 25 + 15?*

Comprehension: *How did you solve the problem?*

Application: *What picture could you draw to show the partial sums?*

Analysis: *How would you compare the partial sums?*

Evaluation: *What is good about the partial sums concept?*

Synthesis: *How would you design a template for*
COMPLEX *reordering partial sums?*

Race to the Top

A game designed to build the concept of partial sums

PLAYERS: 2 OR 3

Materials:

- **Race to the Top Game Board**
- **Four sets of 0-9 Number Cards**
- **Two Pawns**

Directions:

1. **Players shuffle each set of number cards and place cards face down in a pile near each of the template boxes on the game board.**
2. **Players decide the ladder each person will use and place a pawn at the bottom of this ladder.**
3. **The player on the left flips over the top card in the first pile and places it in the first box. He flips over the top card in the second pile and places it in the second box.**
4. **The player on the right does the same with the third and fourth piles of cards and boxes.**
5. **Players read the equation. The goal is not to name the sum. Rather, the goal is to name the partial sums. For example, 36 + 27 = 50 + 13.**
6. **The first player to name the correct partial sums moves his pawn one step up the ladder.** *NOTE: Player 3 is the "judge" determining which player is first to name the partial sums.*
7. **Play continues until the winning player reaches the top of the ladder.** *NOTE: Players can take turns being first to flip cards.*

Content Differentiation:

Moving Back: **Use base ten blocks or other manipulatives to show the partial sums.**

Moving Ahead: **Try this game with 3-digit numbers.**

Race to the Top Game Board

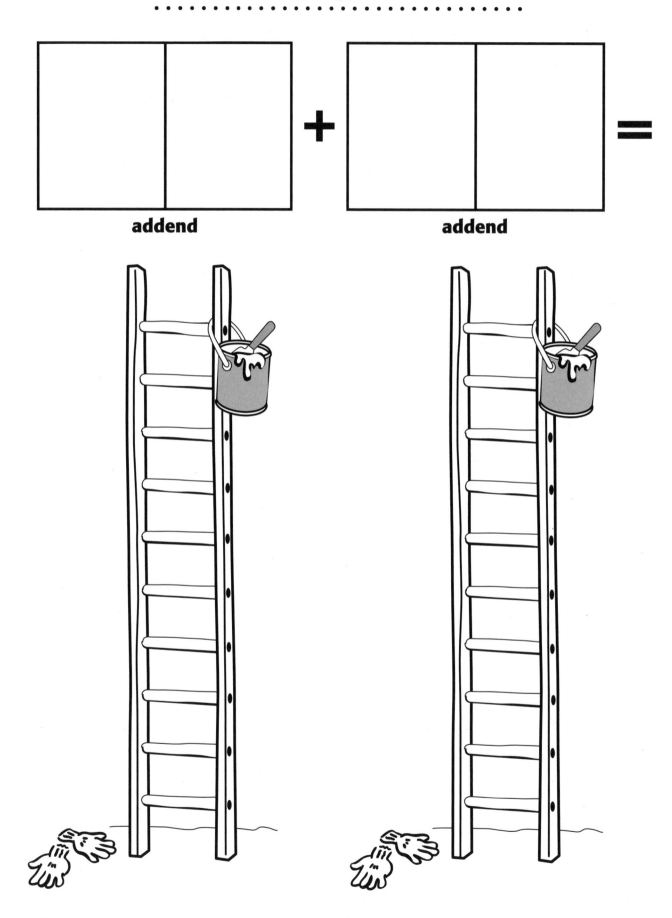

addend   +   addend   =

0-9 Number Cards

0	1
2	3
4	5
6	7
8	9

Partial Differences

 ## What is the Partial Differences Concept?

The partial differences concept is a subtraction strategy that involves combining differences into workable clusters. Like partial sums, this concept serves as an alternative to the traditional algorithm. Instead of following rules that are meaningless to some students, the students using the partial differences strategy do not have to *align columns*, *start with the ones*, or *cross out*. Instead, students find the difference between the tens, then find the difference between the ones, and combine these differences. For example, $57 - 26 = (50 - 20) + (7 - 6)$. In this situation, the difference between the tens is 30 and the difference between the ones is 1. These partial differences are combined into $30 + 1 = 31$. The partial differences concept is also an effective strategy when a subtraction situation involves regrouping. For example, $84 - 36 = (80 - 30) + (4 - 6)$. In this situation, the difference between the tens is 50 and the difference between the ones is -2. These partial differences are combined into $50 - 2 = 48$.

 ## Formative Assessment

To find out if a student understands partial differences, ask the student to solve the following equations;

$37 - 12 =$	$45 - 23 =$	$57 - 34 =$
$64 - 25 =$	$52 - 36 =$	$70 - 28 =$

Ask the student to explain how she solves each problem. If the student uses the traditional algorithm (successfully or not successfully), ask the student if she knows another way to solve the equations. Encourage the student to try the problems without aligning the digits in columns, without starting with the ones, and without crossing out.

Successful Strategies

Using number lines that include negative numbers helps students understand how to find the partial differences when the ones digit value of the minuend is greater than the ones digit value of the subtrahend. It also helps us focus on mathematical truths. It really is a lie when we tell students *you cannot take 6 from 4.* You can take 6 from 4. The result is -2. When the partial differences are combined, you just subtract without the need to memorize rules about adding positive and negative numbers. As with partial sums, even if students successfully use the traditional algorithm, learning the partial differences concept strengthens their conceptual understanding, increases mental math capacity, and improves flexibility and fluency.

Math Words to Use

partial, part, difference, subtract, subtraction, strategy, equation, equal, subtrahend, minuend

Questions at Different *Levels of Cognitive Demand*

EASY

Recall: *What is 46 – 21?*

Comprehension: *How did you solve the problem?*

Application: *What picture could you draw to show partial differences?*

Analysis: *How would you compare the partial differences?*

Evaluation: *What is helpful about the partial differences concept?*

Synthesis: *What if the partial differences were half as many?*

COMPLEX

Race to the Bottom

A game designed to build the concept of partial differences

Materials:

- **Two sets of 0-9 Number Cards** *(see page 128)*
- **One set of 5-9 Number Cards**
- **One set of 1-4 Number Cards**
- **Race to the Bottom Game Board**
- **Number Line and two pawns**

PLAYERS: 2 OR 3

Directions:

1. **Players shuffle each set of number cards and place the cards face down in a pile near each of the template boxes on the game board.** *NOTE: The 5-9 cards are used for the tens digit of the minuend. The 1-4 cards are used for the tens digit of the subtrahend.*
2. **Players decide the ladder each person will use and place a pawn at the top of this ladder.**
3. **The player on the left flips over the top card in the first pile and places it in the first box. He flips over the top card in the second pile and places it in the second box.**
4. **The player on the right does the same with the third and fourth piles of cards and boxes.**
5. **Players read the equation. The goal is not to name the difference. Rather, the goal is to name the partial differences, for example, 63 – 27 = 40 – 4. Use number line as needed.**
6. **The first player to name the correct partial differences moves his pawn one step down the ladder. Play continues until the winning player reaches the bottom of his ladder.** *NOTE: Player 3 is the "judge" determining which player is first to name the partial differences.*

Content Differentiation:

Moving Back: **Use base ten blocks or other manipulatives to show the partial differences.**

Moving Ahead: **Try this game with 3-digit numbers.**

Race to the Bottom Game Board

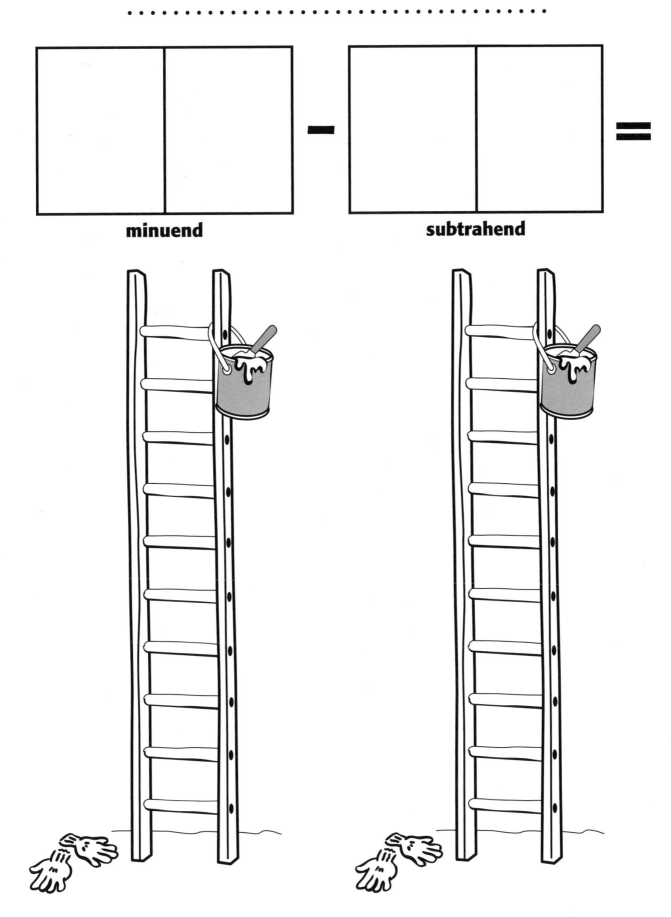

minuend   −   subtrahend   =

Number Line

What is the Near Tens Concept?

The near tens concept involves adjusting numbers based on their proximity to ten. The near tens concept can be used in addition, for example, $57 + 9 = 57 + 10 - 1$ or $57 + 8 = 57 + 10 - 2$. The near tens concept can also be used in subtraction, for example $83 - 9 = 83 - 10 + 1$ or $83 - 8 = 83 - 10 + 2$. When students use the near tens concept, they apply the kind of flexibility with numbers that is needed for mental math and deep conceptual understanding.

 Formative Assessment

To find out if a student understands the near tens concept, ask the student to solve the following equations;

$48 + 10 =$	$56 + 20 =$
$73 - 10 =$	$67 - 30 =$
$53 + 9 =$	$47 + 29 =$
$64 - 9 =$	$93 - 49 =$
$26 + 8 =$	$44 + 28 =$
$56 - 8 =$	$77 - 38 =$

Ask students to explain how they solve each problem. If the student uses the traditional algorithm (successfully or not successfully), ask the student if she knows another way to solve the equations. Invite the student to try the problems using what they know about adding and subtracting tens.

Successful Strategies

To be successful with the near tens concept, it is essential that students have ownership of plus or minus ten. The near tens concept builds upon what students know about adding and subtracting tens. For many students, using the hundred chart helps them to see and apply the near tens concept.

> ### Math Words to Use
>
> **near ten, add, addition, subtract, subtraction, adjust, compensate**

Questions at Different *Levels of Cognitive Demand*

EASY

Recall: *What is 97 – 10 ?*

Comprehension: *How could you use a near ten to solve 43 + 29?*

Application: *How would you show the near tens concept with manipulatives?*

Analysis: *Which problems work better with the near tens concept?*

Evaluation: *When is it not a good plan to use the near tens concept?*

Synthesis: *How would you teach the near tens concept to someone who did not know it?*

COMPLEX

Drop It

A game designed to build the concept of near tens

Materials:

- **Die (labeled 18, 28, 39, 49, 40, 50)**
- **Plus/Minus Spinner, paper clip and pencil**
- **Drop It Hundred Chart**
- **Chip or counter**

PLAYERS: 2

Directions:

1. The first player spins the spinner to find out if he will add or subtract.
2. The first player drops the chip/counter on the hundred chart to find one of the numbers in his equation and rolls the die to find the other number in the equation. *NOTE: The chip/counter is dropped on the 1-50 section (shaded) for addition and the 51-100 section (not shaded) for subtraction. This number serves as the first addend or the minuend. The number on the die serves as the second addend or the subtrahend.* **For example, 74 – 39 = ? The player touches 74 on the hundred chart, moves his finger up four rows of ten (– 40) and forward one space to the right (+ 1). The player describes the near tens situation, for example, 74 – 39 = 74 – 40 + 1.**
3. The sum/difference is the player's positive score for the round.
4. The next player spins the spinner, drops the chip/counter, rolls the die, models, and describes the near tens equation.
5. Players add their scores from each round. The winning player is the player with the highest score after 10 rounds.

Content Differentiation:

Moving Back: Try the game with a die labeled 8, 8, 9, 9, 10, 10.

Moving Ahead: Try the game with a die labeled 38, 48, 58, 69, 79, 89 using a two-hundred chart.

Plus/Minus Spinner

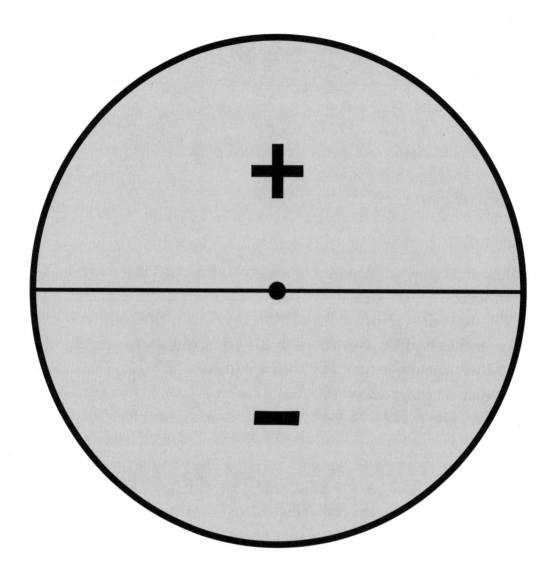

To use the spinner, place a paper clip in the center. Place pencil point through the paper clip at the center of the spinner. Holding the pencil securely with one hand, spin the paper clip with the other hand.

Drop It Hundred Chart

1	2	3	4	5	6	7	8	9	10
11	12	13	14	15	16	17	18	19	20
21	22	23	24	25	26	27	28	29	30
31	32	33	34	35	36	37	38	39	40
41	42	43	44	45	46	47	48	49	50
51	52	53	54	55	56	57	58	59	60
61	62	63	64	65	66	67	68	69	70
71	72	73	74	75	76	77	78	79	80
81	82	83	84	85	86	87	88	89	90
91	92	93	94	95	96	97	98	99	100

What is the Equal Difference Concept?

The equal difference concept is a subtraction strategy based on the idea that different subtraction expressions can have the same difference. The focus is on the value between numbers, for example, $76 - 29 = 77 - 30$. By adding one to the subtrahend and one to the minuend, we create an expression that is more easily solved. In some situations, it makes sense to add two to the subtrahend and the minuend, for example, $52 - 28 = 54 - 30$. In some situations, we may subtract from the subtrahend and the minuend to create a new expression with the same difference. For example, $68 - 41 = 67 - 40$ or $96 - 82 = 94 - 80$. Understanding how to mentally create expressions with the same differences helps students learn to build flexibility and fluency with numbers and operations.

Formative Assessment

To find out if a student understands the equal difference concept, ask the student if the following equations are true or not;

$54 - 29 = 55 - 30$
$76 - 48 = 78 - 50$
$87 - 34 = 90 - 31$
$81 - 34 = 80 - 33$

Ask the student to explain why these equations are true or not true (the third equation is NOT true). Notice if the student solves each side of the equation to verify the answer or if the student uses the equal difference concept.

Successful Strategies

The equal difference concept is applicable in any given subtraction equation, but it is most efficient when the created expression invites mental math. For example, $73 - 25 = 74 - 26$. The created expression does not necessarily increase opportunities for mental math. It is beneficial for students to think about when to use the equal difference concept and when it may be better to use a different concept.

Math Words to Use

equal, difference, subtract, subtraction, add, addition, expression, equation

---

 Questions at Different *Levels of Cognitive Demand*

EASY

Recall: *What is the difference between 52 and 39?*

Comprehension: *Which examples could you give of the equal difference concept?*

Application: *How would you show the equal difference concept on a number line?*

Analysis: *How are these expressions related?*

Evaluation: *Which problems work best with the equal difference concept?*

Synthesis: *What if you were to create a new way to record twice the difference?*

COMPLEX

All Strung Out

*A game designed to build the concept
of equal difference*

**PLAYERS:
2 OR MORE**

Materials:

- **Number Cards (50-98)**
- **One Subtrahend Spinner (+2 or +1 or –1)**
- **Paper clip and pencil**
- **Number line (1-100)** *NOTE: If number line is not available, cut apart the rows of a hundred chart and tape the ends together.*
- **String/yarn, scissors**

Directions:

1. Shuffle the number cards and place face-down in a pile. The first player draws a card. This number will be the minuend.
2. The first player spins the spinner to find the subtrahend for the subtraction expression. This player cuts a piece of string that shows the difference (distance between the minuend and the subtrahend). The player moves the string two numbers to the right (when using the +2 spinner) and describes the equal difference equation using the term "is the same as." For example, 86 – 28 *is the same as* 88 – 30.
3. If the player accurately models and describes the equal difference equation, the player keeps the string.
4. Players take turns drawing number cards, spinning, modeling, and describing the equal difference expressions.
5. After each player has five turns, all players line up their pieces of string, and compare their total lengths. The winner is the player with the most string.

Content Differentiation:

Moving Back: Use the +1 spinner (move string <u>one</u> number to the right).

Moving Ahead: Use the –1 spinner (move string <u>one</u> number to the left) or make your own spinner.

All Strung Out Number Cards

50	51	52	53	54
55	56	57	58	59
60	61	62	63	64
65	66	67	68	69
70	71	72	73	74

All Strung Out Number Cards

75	76	77	78	79
80	81	82	83	84
85	86	87	88	89
90	91	92	93	94
95	96	97	98	

Subtrahend Spinner Options

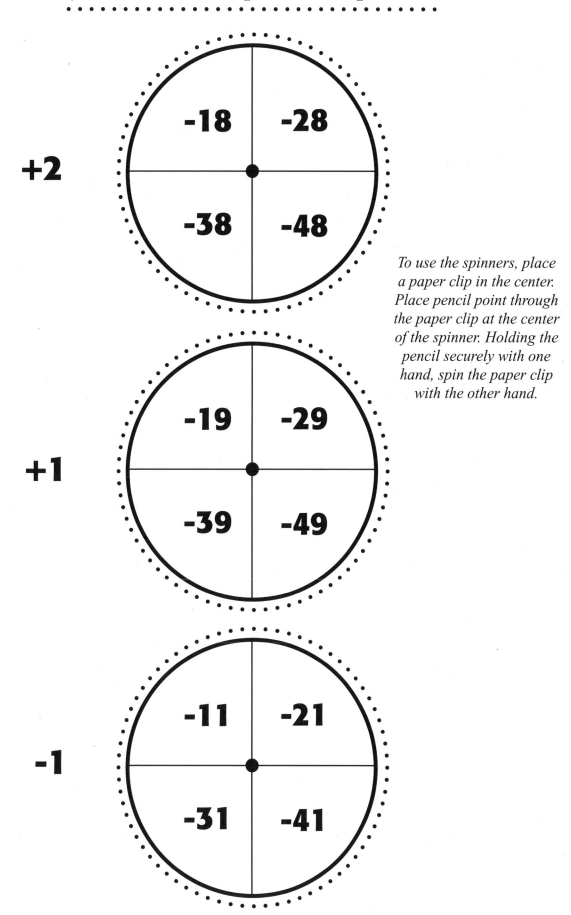

+2

-18 | -28

-38 | -48

To use the spinners, place a paper clip in the center. Place pencil point through the paper clip at the center of the spinner. Holding the pencil securely with one hand, spin the paper clip with the other hand.

+1

-19 | -29

-39 | -49

-1

-11 | -21

-31 | -41

Notes

Notes

Notes

Notes

Notes

References

Bermejo, V. (1996). Cardinality Development and Counting. *Developmental Psychology*, vol. 32, no. 2, pp 263-268.

Bloom, B., Englehart, M. Furst, E., Hill, W., & Krathwohl, D. (1956). Taxonomy of Educational Objectives; The Classification of Educational Goals. *Handbook I: Cognitive Domain*. New York, Toronto: Longmans Green.

Clements, D. (1999). Subitizing: What is it? Why teach it? *Teaching Children Mathematics*, 5 (March): 400-405.

Freeman, F. N. (1912). Grouped Objects as a Concrete Basis for the Number Idea. *Elementary School Teacher* 8, 306-314.

Kaufman, E. L., Lord, M.W., Reese, T. W. & Volkman, J. (1949). The Discrimination of Visual Number. *American Journal of Psychology*, 62, 498-525.

McIntosh, A. J., Reyes, B. J., and Reys, R. E. (1993). A Proposed Framework for Examining Number Sense. *For the Learning of Mathematics*, 12 (3), 2-8.

National Association for the Education of Young Children and National Council of Teachers of Mathematics. 2002. *Early Childhood Mathematics: Promoting Good Beginnings, A Joint Position Statement*. Reston, VA: NCTM.

National Council of Teachers of Mathematics (NCTM). *Curriculum Focal Points for Prekindergarten Through Grade 8 Mathematics: A Quest for Coherence*. Reston, VA: NCTM, 2006.

National Council of Teachers of Mathematics (NCTM). *Principles and Standards for School Mathematics*. Reston, VA: NCTM, 2000.

National Mathematics Advisory Panel (2008). *Foundations for Success: The Final Report of the National Mathematics Advisory Panel*, U.S. Department of Education: Washington, DC.

OED2, *Oxford English Dictionary, Second Edition Online*. New York, New York: Oxford University Press.

Piaget, J. (1964). Development and Learning. In R. Ripple & V. Rockcastle (Eds.). *Piaget Rediscovered*. Ithaca New York: Cornell University Press.

Piaget, J. (1970). Piaget's Theory. In P.H. Mussen (Ed.). *Carmichael's Manual of Child Psychology*, vol. 1, 3rd edition. London: John Wiley and Sons, Ins.

Taylor-Cox, J. (2001). How Many Marbles in the Jar? Estimation in the Early Grades. *Teaching Children Mathematics*, 8 (4), 208-2 14.

Taylor-Cox, J. (2005). *Family Math Night: Math Standards in Action*. Larchmont, NY: Eye on Education.

Taylor-Cox, J. & Kelly, B. (2008). The Doctors Disagree: Two Points of View on the Topic of "Drill and Kill" in Mathematics Education. *The Banneker Banner; The Official Journal of the Maryland Council of Teachers of Mathematics*, 25 (1), 6-7.